D1199331

Algebra
and Geometry
for Teachers

Charles F. Brumfiel
University of Michigan

Irvin E. Vance
Michigan State University

Addison-Wesley Publishing Company

Reading, Massachusetts
Menlo Park, California · London · Don Mills, Ontario

This book is in the
Addison-Wesley Series in Mathematics

Preface

The primary purpose of this book is to provide an integrated course in algebra and geometry for the training or retraining of elementary teachers. The coverage of topics is approximately that recommended by the Committee on the Undergraduate Program in Mathematics (CUPM).

The level of sophistication gradually increases throughout the text. Chapter 1, "Intuitive Algebra," provides a review of basic concepts, but its primary purpose is to establish algebra as a useful language. Unhappily, to many students algebra is a collection of formal, relatively meaningless skills. This chapter offers material to help re-orient these negative attitudes.

The early geometry chapters deal with the concepts that are central to the geometry program of the elementary grades, the concepts of construction, congruence, measurement, parallelism, and perpendicularity. In these chapters the student should learn by doing. The presentation is informal. Many of the important ideas are developed in sequences of problems. A semi-programmed style of writing has been adopted.

We return to algebra in two chapters which stress structure. We use this opportunity to review properties of number systems. Finite fields provide a beautiful and relatively simple setting in which students can take a fresh look at solving equations, factoring polynomials, graphing functions, and organizing informal proofs. A single semester course taught for elementary teachers might terminate here or at most include a few topics from those later chapters that the instructor considers most appropriate for his class.

Actually, the text from Chapter 5 on can be used effectively as a basic text for a mathematics course for Junior High School teachers. Our chief goal in these later chapters is to bring geometric and algebraic ideas together through the unifying concept of a group of transformations. Chapters on logic, the axiomatic structure of geometry, and analytic geometry complete the text.

The authors have used the material presented here in both pre-service and in-service teacher training courses at Michigan State University and the University of Michigan. The reaction of students has been encouraging. The great variety of examples in the problem lists help to maintain student interest.

We wish to thank the entire staff at Addison-Wesley for the care with which they have processed the manuscript.

December, 1969   C. F. B.

I. E. V.

Contents

1–1 Introduction

The main purpose of this chapter is to provide an opportunity for the reader to review basic algebraic concepts. The emphasis will be upon informal algebraic reasoning. For simplicity, most computation will be restricted to whole numbers and integers. We assume that the reader is familiar with elementary concepts of set theory. But even if this is not the case, most of the set theoretic notation and terminology used in the text should be understood. The exercise list below reviews several key ideas and the symbolism of set theory.

Exercises 1–1

1. We denote the set of whole numbers by W:

$$W = \{0, 1, 2, 3, \ldots\}.$$

The use of the three dots reflects our confidence in your ability to continue the counting process indefinitely. Translate each symbolic statement below and classify it as true or false.

a) $7 \in W$ b) $\frac{1}{2} \notin W$ c) $0 \notin W$

d) $\{1, 2, 3\} \subset W$ e) $\{1, -1\} \not\subset W$

f) $\{1, 2, 3\} \cup \{2, 3, 4\} = \{1, 2, 3, 4\}$

g) $\{1, 2, 3\} \cap \{2, 3, 4\} = \{2, 3\}$

h) $\{x \in W \mid x < 5\} = \{0, 1, 2, 3, 4, 5\}$

(Read the left member as "The set of all x in W such that x is less than 5.")

i) $\{x \in W \mid 3 \le x < 7\} = \{3, 4, 5, 6\}$

j) If $N = \{x \in W \mid x \ne 0\}$, then $N \cup \{0\} = W$.

k) $\{3\} \neq 3$ l) $\{0\} = 0$ m) $\{1, 2, 3\} \cap \{4, 5\} = \emptyset$

n) $\{x \in W \mid x < 0\} = \emptyset$

2. We denote the set of integers by I:

$$I = \{\ldots, -2, -1, 0, 1, 2, \ldots\}.$$

If a and b are whole numbers and $a > b$, how do the negative integers $-a$ and $-b$ compare?

3. Translate each statement below and decide whether it is true or false.

   a) $W \subset I$ *   b) $I \supset W$

   c) $W = \{x \in I \mid x > -1\}$

   d) The complement of W in I is the set, $\{x \in I \mid x \notin W\}$.

   e) $W \cap I = W$

4. Let

$$E = \{x \in I \mid x = 2y \text{ for some } y \in I\},$$

and let

$$O = \{x \in I \mid x = 2y + 1 \text{ for some } y \in I\}.$$

   a) What are the common names for sets E and O?

   b) Describe the sets $E \cup O$ and $E \cap O$.

5. Let E and O be the sets of Example 4. Determine the truth or falsity of each statement below.

   a) $\forall x, y \in E, x + y \in E$.

   (Read the symbol $\forall$ as "for all.")

   b) $\forall x, y \in O, x + y \in O$.

   c) $\forall x \in E, \forall y \in O, x + y \in O$.

   d) $\forall x \in E, \forall y \in O, x \cdot y \in O$.

   e) $\exists x \in E$ such that $\forall y \in I, y + x = y$.

   (Read the symbol $\exists x$ as "there exists an x.")

   f) $\exists x \in O$ such that $\forall y \in I, y \cdot x = y$.

   g) $x \in I, x \notin O \Rightarrow x \in E$.

   (Read the symbol $\Rightarrow$ as "implies" or read this sentence as, "If $x \in I$ and $x \notin O$, then $x \in E$.")

   h) $\forall x, y \in E, x \cdot y = x \Rightarrow x = 0$.

---

* We use the symbol $\subset$ in the sense that some authors use $\subseteq$. Thus $W \subset W$ is a true statement.

i) $\forall x, y \in O, x \cdot y = x \Rightarrow y = 1.$

j) $\forall x, y \in I, x \cdot y = x \Rightarrow x = 0$ or $y = 1.$

k) $\nexists x \in O$ such that $x \cdot x \in E.$

l) $\nexists x \in E$ such that $x \cdot x \in E.$

m) $\forall x, y \in O, xy - y \in E.$

n) $\forall x, y \in O, x - y \in O.$

o) $\forall x, y \in O, x < y \Rightarrow \exists z \in E$ such that $x < z < y.$

The preceding exercises make free use of *variables*. Loosely speaking, variables are *meaningless* symbols which can be replaced by certain *meaningful* symbols. The usual variables are letters. Suppose that in the sentence,

x is even,

it is understood that the symbol x can be replaced by any symbol naming a whole number. Then from the sentence "x is even," we can form the infinite collection of statements:

0 is even; 1 is even, 2 is even, . . .

by making all permissible replacements of the variable x. Note that some of these statements are true,

0 is even, 2 is even, . . .

and some are false,

1 is even, 3 is even, ...

The original expression, x is even, is called an *open sentence*. Selection of the term *open* is motivated by the fact that the sentence is neither true nor false. Its truth is open until a specific number replaces the variable.

The idea of a variable is perhaps brought out most clearly by *placeholder* notation. This notation is used in the early elementary grades. Instead of letters for variables one employs blanks or geometric shapes in which numbers may be written.

a) ____ is even corresponds to x is even.

b) $\square$ is a prime number corresponds to x is a prime number.

c) $\triangle + \triangle = 2 \cdot \triangle$ corresponds to $x + x = 2 \cdot x.$

d) $\triangle + \square = \bigcirc$ corresponds to $x + y = z.$

Different placeholders play the role of different letters. We shall use this notation briefly in the next few pages.

Exercises 1–2

In the following exercises let the replacement set for all variables and placeholders be the set,

$$\{x \in W \mid x \le 10\}.$$

1. How many statements can be formed from the open sentence $y \le 5$ by permisssible replacements of the variable?

2. In Exercise 1 how many of the resulting statements are true, and how many are false?

3. A student "solves" the equation

$$x + x = 8$$

by writing $5 + 3 = 8$, and says, "The first x can be 5 and the second x, 3." This is a violation of our agreement on the replacement of variables. Explain.

4. In solving the equation $\square + \square = 8$, a student writes

$$\boxed{6} + \boxed{2} = 8.$$

Discuss this abuse of the use of placeholders.

5. A student asserts that the open sentence $x + y = 4$ is true *only* for the pairs of whole numbers

$$(0, 4), (1, 3), (3, 1), (4, 0).$$

He claims that since the variables x and y are different, they must be replaced by different numbers. Discuss.

6. A student, solving the equation $\triangle + \square = 4$, writes out the following solutions and thinks he has listed all possibilities.

$$\triangle\!\!\!\!_{0} + \boxed{4} = 4, \qquad \triangle\!\!\!\!_{1} + \boxed{3} = 4,$$
$$\triangle\!\!\!\!_{3} + \boxed{1} = 4, \qquad \triangle\!\!\!\!_{4} + \boxed{0} = 4.$$

Discuss his misunderstanding of the symbolism.

7. In how many different ways can the placeholders be filled from our replacement set of eleven elements so that the resulting statement is true?

a) $\square + 3 = 8$

b) $\square + 3 = 15$

c) $\square + \square = 8$

d) $\square + \triangle = 8$

e) $\square + \square = 2 \times \square$

f) $\square + \bigcirc = \bigcirc + \square$

g) $\square + \triangle = \bigcirc$

h) $\square + \triangle < 7$

i) $\square > \triangle + 8$

j) $\square = (2 \times \triangle) + 3$

1–2 Algebraic Reasoning

One tends to think of arithmetical activity as the performance of operations upon numbers. One is given the numbers and told essentially what to do to them. There is limited scope for ingenuity. The reasoning required for problems that we classify as algebraic is generally more devious. No clear line can be drawn between arithmetical and algebraic reasoning, of course, but in algebra one is concerned primarily with *generalizations* and with *properties* of the operations that are employed, while in arithmetic one's attention is fastened chiefly upon numerical computation. There is much more "if ... then ..." reasoning in algebra than there is in arithmetic. The problem below is a typical problem of arithmetic.

> One hundred tickets were sold for a school play. Of these, 60 were student tickets and the rest adult. A student ticket sells for 50¢ and an adult ticket for 75¢. Find the total receipts.

Solving this problem is a matter of performing operations of subtraction, multiplication, and addition upon the numbers 100, 60, 50, and 75. With modifications we get an algebraic problem.

> One hundred tickets were sold for a school play. Adult tickets sold for 75¢ and student tickets for 50¢. If total receipts were $60, how many of each were sold?

A child who knows nothing about formal algebraic techniques can still solve this problem by the following "algebraic" type of reasoning:

1. *If* all tickets sold had been student tickets, *then* receipts would have been only $50. Actual receipts were $10 more than this, so some adults must have bought tickets.

2. Since an adult ticket sells for 25¢ more than a student ticket, 4 adult tickets bring in $1 more than do 4 student tickets.

3. To get $10 more, then, 10 × 4 adult tickets must have been sold. So 40 adult and 60 student tickets were sold.

   An arithmetical *formula* for the first problem is

   $$(50 \times 60) + 75 \times (100 - 60).$$

An algebraic *equation* for the second problem is

$$75y + 50(100 - y) = 6000.$$

Traditionally, the study of algebra has stressed techniques for setting up and solving equations. We are using the phrase, "algebraic reasoning," in a rather vague sense to refer to ways of thinking that lead to the solution of problems of this sort. In the exercise list below we formulate problems and suggest various ways of thinking "algebraically."

Exercises 1–3

1. Twenty cards are placed in two piles so that one pile contains 6 more than the other. How many are in each pile?

   [*Hint:* Put ten cards in each pile. If one card is transferred from one pile to the other, how many more are in one pile than the other? How many must be moved to make a difference of 6?]

2. There are 32 animals, chickens and pigs, in a barnyard. These animals have a total of 84 legs. How many chickens are there?

   [*Hint:* If they were all chickens, there would be only 64 legs. How many more legs are needed? How many chickens must be exchanged for pigs?]

3. A ring and watch together cost $140. The watch cost $40 more than the ring. How much did each cost?

   [*Hint:* If the watch had been $40 cheaper, the total cost would have been only $100. How much would each have cost in this case?]

4. Joe is 4 years older than Jim. In 5 years the sum of their ages will be 32. How old is each now?

   [*Hint:* If Joe were 4 years younger than he is, both boys would be the same age and the sum of their ages in 5 years would be 28 instead of 32.]

5. One number is twice as large as a second. It is also 7 more than the second. Determine the numbers.

   [*Hint:* Try 5 for the little number. Doubling, one gets 10. This doesn't quite work.]

6. A loaf of bread cost 10¢ more than a quart of milk. For 3 loaves of bread and 2 quarts of milk I pay $1.35. What is the price of each?

   [*Hint:* If I had bought 5 loaves of bread and no milk, instead of 3 loaves of bread and 2 quarts of milk, what would the cost have been?]

7. On a balance scale, 5 large weights and 3 little ones just balance 3 large weights and 15 little ones. How many small weights balance one large one?

   [*Hint:* Remove some weights from both sides, keeping the scale in balance.]

8. Find a number such that when it is added to 30 the sum is three times as much as the difference when it is subtracted from 30.

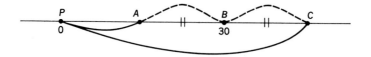

   [*Hint:* Visualize the number line above. How do segments *PA* and *PC* compare?]

9. Modify Exercise 8 by requiring that the sum be five times as much as the difference. Explain why a diagram of this type is not so helpful if the sum is to be four times the difference.

10. We can use the symbols b and m to represent a loaf of bread and a quart of milk, respectively. Let us agree that the expression

$$b, b, b, m, m \rightarrow \$1.30$$

indicates that three loaves of bread and two quarts of milk cost $1.30. Suppose that we know also that

$$b, b, m, m, m \rightarrow \$1.25.$$

a) Which costs more, a loaf of bread or a quart of milk?

b) How much more?

c) What is the cost of one loaf of bread and four quarts of milk?

d) What is the cost of five quarts of milk?

e) What is the cost of one quart of milk? Of one loaf of bread?

11. Work the bread and milk problems sketched below using the type of "visual" reasoning suggested.

a) $b, b, m \rightarrow 95¢$
$b, m \rightarrow 60¢$
$b \rightarrow r$
$m \rightarrow s$

b) $b, b, m, m \rightarrow \$1.30$
$b, m, m, m \rightarrow 1.21$
$b, m \rightarrow r$
$m, m \rightarrow s$
$m \rightarrow t$
$b \rightarrow u$

c) $b, b, b, m, m \rightarrow \$1.30$
$b, b, m, m, m \rightarrow 1.20$
$b, b, b, b, b, m, m, m, m, m \rightarrow r$
$b, m \rightarrow s$
$b \rightarrow u$
$m \rightarrow v$

1–3 Equations and Inequalities

In this section we shall consider equations and inequalities. We are not interested in the development of formal techniques. In all the problems of this section emphasis is placed upon the drawing of *inferences*. For example, you will be asked to examine a set of equations and produce new equations which you will deduce from the given ones. We shall not attempt to analyze the logical processes by which you will make your deductions. You are urged not to rely upon mechanical skills that you may have retained from your training in school mathematics. Instead, bring an intuitive, common-sense approach to the work. Whereas the reasoning that was required of you in the preceding section was highly verbal, the problems of this chapter stress symbolic, nonverbal reasoning. Many of the problems of this chapter are written in a programmed style, directing your thinking along particular lines. Throughout this entire section the reference set for all variables will be the set W. That is, an equation like $2x = 1$ is off bounds for us. It has no solution in our set. Also, an equation like $x^2 = 25$ will be considered to have only the single solution 5. We shall ignore the root -5 which lies in the set I.

Exercises 1–4

1. Copy the equations below, replacing each of the letters $a, b, \ldots$ by the appropriate whole number.

    a) $x + 5 = 8$
       $x = a$

    b) $x - 4 = 7$
       $x = a$

    c) $7 - x = 2$
       $x = a$

    d) $3x = 12$
       $x = a$

    e) $\dfrac{24}{x} = 8$
       $x = a$

    f) $7x = 0$
       $x = a$

    g) $2x + 4 = 22$
       $2x = a$
       $x = b$

    h) $2(x + 3) = 30$
       $x + 3 = a$
       $x = b$

    i) $2x + 10 = 18$
       $x + 5 = a$
       $x = b$

    j) $\dfrac{x}{2} - 3 = 4$
       $\dfrac{x}{2} = a$
       $x = b$

    k) $\dfrac{36}{x} - 3 = 6$
       $\dfrac{36}{x} = a$
       $x = b$

    l) $\dfrac{24}{7 - x} = 8$
       $7 - x = a$
       $x = b$

2. Use the given information and draw the suggested inferences.

    a) If $x + y = 7$, then $2x + 2y = a$.
    b) If $x + y + 4 = 15$, then $x + y - 3 = a$.
    c) If $x + y + 2 = 11$, then $2x + 2y + 4 = a$.
    d) If $x + y + 2 = 10$, then $2x + 2y - 3 = a$.
    e) If $y - (x + 3) = 7$, then $y - x = a$.
    f) If $y - (x + 3) = 7$, then $y - (x - 2) = a$.
    g) If $2y - 4x + 6 = 20$, then $y - 2x + 5 = a$.
    h) If $x + 2y - z = 5$, then $3x + 6y - 3z + 4 = a$.
    i) If $5x = 3x + 8$, then $2x = a$.
    j) If $5x + 3 = 4x + 7$, then $x = a$.
    k) If $5x - 10 = 4x - 6$, then $x = a$.

l) If $5x + 3y = 3x + 8y$, then $2x = a$.

m) If $5x - 7y = 2x - 2y$, then $3x = a$.

n) If $xy = x$ and $x = 3$, then $y = a$.

o) If $xy = y$ and $x = 3$, then $y = a$.

p) If $xy = x$ and $x \neq 0$, then $y = a$.

q) If $xy = x$ and $y \neq 1$, then $x = a$.

r) If $xy = x$, then either $x = a$ or $y = b$.

s) If $xy = 0$, then either $x = a$ or $y = b$.

t) If $(x - 2)(y - 4) = 0$ and $x \neq 2$, then $y = a$.

u) If $(x - 4)(y - 7) = 0$, then $x = a$ or $y = b$.

v) If $y(x - 3) = y$, then $x = a$ or $y = b$.

w) If $\dfrac{x - 3}{y} = x - 3$, then $x = a$ or $y = b$.

3. Follow the suggested steps and solve the following systems of equations.

a) $\begin{cases} x + (x + y + 2) = 10 \\ \qquad x + y + 2 = 7 \end{cases}$
$$x + a = 10$$
$$x = b$$
$$y = c$$

b) $\begin{cases} 2x + (x + y + 4) = 20 \\ \qquad x + y = 6 \end{cases}$
$$2x + a = 20$$
$$x = b$$
$$y = c$$

c) $\begin{cases} 3x + y = 14 \\ 2x + y = 11 \end{cases}$
$$x = a$$
$$y = b$$

d) $\begin{cases} 2x + y + 4 = 10 \\ \ x + y + 4 = 7 \end{cases}$
$$x = a$$
$$y = b$$

e) $\begin{cases} 3x + 4y = 11 \\ 3x + 6y = 15 \end{cases}$
$$2y = a$$
$$y = b$$
$$x = c$$

f) $\begin{cases} 2x - 3y = 10 \\ 2x - 6y = 4 \end{cases}$
$$3y = a$$
$$y = b$$
$$x = c$$

g) $\begin{cases} 2x + 3y = 23 \\ 4x + 7y = 51 \end{cases}$
$$4x + 6y = a$$
$$y = b$$
$$x = c$$

h) $\begin{cases} 3x - 2y = 12 \\ \ 6x + y = 39 \end{cases}$
$$6x - 4y = a$$
$$5y = b$$
$$y = c$$
$$x = d$$

4. The problems below raise questions about inequalities. Draw the inferences that seem most interesting.

a) If $x + 4 > 7$, $x + 10 > a$.

b) If $x > 3$, $4x > a$.

c) If $x + 2 < 11$, $2x - 3 < a$.

d) If $2x - 3 < 21$, $x + 5 < a$.

e) If $5 < x < 10$, $a < 20 - x < b$.

f) If $x + y + 3 > x + 2y$, then $y < a$.

g) If $x + 1 > 7$ and $y + 2 > 9$, then $x + y + 3 > a$.

5. Assuming that each inequality on the left is a true statement, indicate which of the two alternatives on the right must hold.

a) $x - y > x - 7$, $y > 7$, $y < 7$

b) $24 - x > y - x$, $y > 24$, $24 > y$

c) $2x + y < 2y + x$, $x < y$, $y < x$

d) $2x - y < 2y - x$, $x < y$, $y < x$

e) $x + y + 9 > x + 2y$, $y > 9$, $y < 9$

f) $x - y > x - 2y + 9$, $y > 9$, $y < 9$

6. Each of the problems below has two whole number solutions. You are expected to find these by trial and error. However, before completing the list you should discover an interesting pattern that will simplify your task.

Example. $x^2 + 6 = 5x$.

The solution set is 2, 3, for

$$2^2 + 6 = 5 \cdot 2 \qquad \text{and} \qquad 3^2 + 6 = 5 \cdot 3,$$

and no other whole number satisfies the equation.

a) $x^2 + 3 = 4x$ b) $x^2 + 8 = 6x$

c) $x^2 + 10 = 7x$ d) $x^2 + 12 = 7x$

e) $x^2 + 6 = 7x$ f) $x^2 + 30 = 11x$

g) $x^2 + 40 = 13x$ h) $x^2 + 300 = 103x$

i) $x^2 + 150 = 25x$ j) $x^2 + 7000 = 1007x$

k) $x^2 + 156 = 25x$ l) $x^2 + 154 = 25x$

m) $x^2 + 136 = 25x$ n) $x^2 + 46 = 25x$

7. One value of x satisfying the equation

$$x^2 + a = 134x$$

is the number 7. What is the other solution?

8. One value of x satisfying the equation

$$x^2 + 1400 = ax$$

is the number 7. What is the other solution?

When we interpret numbers as lengths, certain products have simple geometric interpretations. The diagrams below illustrate this.

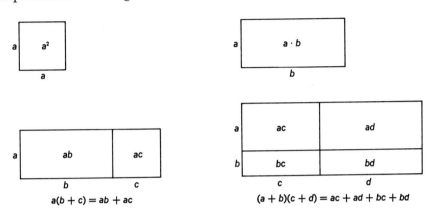

$$a(b + c) = ab + ac$$

$$(a + b)(c + d) = ac + ad + bc + bd$$

Exercises 1–5

1. Write the formula suggested by each diagram below.

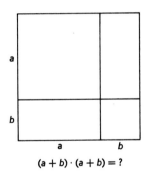

$$(a + b) \cdot (a + b) = ?$$

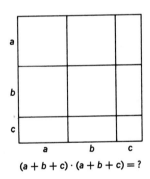

$$(a + b + c) \cdot (a + b + c) = ?$$

2. Explain how the diagram shows that the sum of all the numbers in the box is $(1 + 2 + 3 + 4)^2$.

	1	2	3	4
1	1	2	3	4
2	2	4	6	8
3	3	6	9	12
4	4	8	12	16

3. Explain how the diagrams show that

$$2^2 = 1 + 2 + 1, \qquad 3^2 = 1 + 2 + 3 + 2 + 1,$$
$$4^2 = 1 + 2 + 3 + 4 + 3 + 2 + 1.$$

  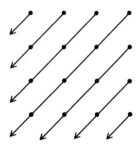

4. Use Exercise 3 and explain how the diagram shows that the sum of the numbers in the array is $1^2 + 2^2 + 3^2 + 4^2$.

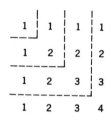

5. Use Exercise 4 and explain how the diagram shows that the sum of the numbers in the box is $1^3 + 2^3 + 3^3 + 4^3$.

1	2	3	4
2	4	6	8
3	6	9	12
4	8	12	16

```
1 X  1 ┊ 1 ┊ 1 ┊ 1
─────────┘   ┊   ┊
2 X  1    2 ┊ 2 ┊ 2
────────────┘   ┊
3 X  1    2    3 ┊ 3
────────────────┘
4 X  1    2    3    4
```

6. Combine the results of Exercises 2 and 5 and show that

$$(1 + 2 + 3 + 4)^2 = 1^3 + 2^3 + 3^3 + 4^3.$$

Show similarly that

$$(1 + 2 + 3 + 4 + 5 + 6)^2 = 1^3 + 2^3 + 3^3 + 4^3 + 5^3 + 6^3.$$

Suggest a general formula.

7. Using the diagrams below, show that

$$1 + 2 + 3 + \cdots + 10 = 5 \cdot 11$$

and

$$1 + 2 + 3 + \cdots + 100 = 50 \cdot 101$$

$$1 + 2 + 3 + \cdots + 8 + 9 + 10$$

$$1 + 2 + 3 + \cdots + 98 + 99 + 100$$

8. Given that n is an even whole number, devise a formula for the sum of the n numbers below.

$$1 + 2 + 3 + \cdots + (n - 2) + (n - 1) + n.$$

9. Given that n is an odd whole number, devise a formula for the sum

$$1 + 2 + \cdots + (n - 1) + n.$$

10. The product $(a + b)(c + d + e)$ can be written as the sum of 6 numbers,

$$ac + ad + ae + bc + bd + be.$$

For each product below, what is the number of terms in the sum if the indicated multiplications are performed?

a) $(a + b)(c + d)$

b) $(a + b + c)(d + e + f)$

c) $(1 + 2 + 2^2)(1 + 3)$

d) $(1 + 2)(1 + 3)(1 + 5)$

e) $(1 + 2 + 2^2 + 2^3)(1 + 5 + 5^2 + 5^3)$

f) $(1 + 2 + 2^2 + 2^3)(1 + 3)(1 + 5)$

The techniques of Exercise 10 in the above list are useful in obtaining information about the factors of a number. For example, note that $6 = 2 \cdot 3$ and the factors of 6 are

$$1, 2, 3, 2 \cdot 3.$$

These factors are obtained by multiplying out the expression below:

$$(1 + 2)(1 + 3) = 1 + 3 + 2 + 2 \cdot 3.$$

Similarly, one gets all factors of

$$12 = 2^2 \cdot 3$$

by multiplying out

$$(1 + 2 + 2^2)(1 + 3) = 1 + 2 + 2^2 + 3 + 2 \cdot 3 + 2^2 \cdot 3.$$

Inspecting this last expression,

$$(1 + 2 + 2^2)(1 + 3) = 7 \cdot 4,$$

one sees

  i) 12 has six factors. (Why?)

  ii) The sum of all factors of 12 is $7 \cdot 4 = 28$. (Why?)

Since $30 = 2 \cdot 3 \cdot 5$ we can compute the number of factors of 30 and the sum of all these factors as indicated below. Explain.

$$30 = 2 \cdot 3 \cdot 5 \to (1 + 2)(1 + 3)(1 + 5) \to 30 \text{ has 8 factors.}$$
$$(1 + 2)(1 + 3)(1 + 5) = 3 \cdot 4 \cdot 6 \to 72 \text{ is the sum of all factors of 30.}$$

Problem. Use the fact that $1000 = 2^3 \cdot 5^3$ and show from the expression

$$(1 + 2 + 2^2 + 2^3)(1 + 5 + 5^2 + 5^3)$$

that the number 1000 has 16 factors, and the sum of all these factors is 2340.

How many factors has 1,000,000? What is the sum of all these factors? Show that the sum of all the factors of 120 is 360 (exactly $3 \cdot 120$). Try to find a number such that the sum of all its factors is at least 4 times the number itself. Find a number which has exactly 1000 factors; exactly 1001 factors. Show that every number which has an odd number of factors is a perfect square.

1–4 Absolute Value

In the set of integers it is convenient to associate each *integer* x with a *whole number* called the *absolute value* of x and denoted by $|x|$. The absolute value of each whole number is just that number itself

if $x \geq 0$,   then $|x| = x$;

$|0| = 0$, $|3| = 3$, $|7| = 7$, ...

For each negative integer x the absolute value is that whole number which is the negative of x.

If $x < 0$,   then $|x| = -x$;

$|-2| = -(-2) = 2$, $|-7| = 7$, ...

Perhaps the most convenient way to think about absolute value is geometrically. The absolute value of any integer can be thought of as its distance from the zero point when we represent the integers in the usual way as points on a line.

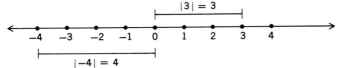

The concept of absolute value is a very useful tool and lends itself to the construction of many interesting problems.

$$|7 - 4| = 3, \qquad |4 - 7| = 3.$$

In each case the absolute value of the difference can be interpreted as the distance between the two points 4 and 7 on the number line. In general, if $a \neq b$, each of

$$|a - b| \qquad \text{and} \qquad |b - a|$$

denotes the positive number that is the length of the line segment with a and b as endpoints on the number line:

$|8 - 2| = 6$, the length of the segment from 2 to 8,

$|8 + 2| = |8 - (-2)| = 10$, the length of the segment from -2 to 8,

$|-3 - 11| = 14$, the length of the segment from 11 to -3,

$|2 + 11| = |2 - (-11)| = 13$, the length of the segment from -11 to 2.

Inequalities which involve absolute value are of particular importance. Note that the equation

$$|x - 3| < 5$$

is satisfied by all integers *less than 5 units distance from* 3.

We can visualize this set of integers as a set of points centered about 3. For the inequality

$$|x - 3| \leq 5$$

the points -2 and 8 would be included. For the equation

$$|x - 3| = 5,$$

the two numbers 8 and -2, one 5 units to the right and the other 5 units to the left of 3, satisfy the equation. Many problems involving absolute value can be solved by visualizing the problems geometrically.

Exercises 1–6

1. Give the absolute value for each of the following:
   a) $|-4|$ b) $|4|$ c) $|7 - 2|$ d) $|2 - 7|$

2. Give two values of x satisfying each equation.
   a) $|x| = 3$ b) $|x| = 4$ c) $|x| = 11$

3. Give each absolute value by visualizing the distance between two points on the number line.
   a) $|8 - 3|$ b) $|2 - 9|$ c) $|4 - (-5)|$
   d) $|-2 - 6|$ e) $|-2 - (-7)|$ f) $|-15 - (-4)|$
   g) $|-2 + 3| = |-2 - (-3)|$ h) $|-3 + 5|$

4. Give the two values of x satisfying each equation below by visualizing two points on the number line equidistant from a certain point.

   Example. $|x - 3| = 7$. The point x may be 7 units to the right or left of 3. Hence $x = 10$ or -4.

   a) $|x - 2| = 5$ b) $|x - 7| = 2$ c) $|x - (-4)| = 4$
   d) $|x + 2| = 9$ [*Hint:* Write as $|x - (-2)| = 9$.]
   e) $|x + 4| = 7$ f) $|x + 4| = 3$

5. It is intuitively clear that
   $$|2x - 4| = 2 \cdot |x - 2|; \qquad |3x + 15| = 3 \cdot |x + 5|, \quad \text{etc.}$$
   Use this concept and the technique of Exercise 4 to solve the equations below.
   a) $|2x - 10| = 14$ b) $|3x + 21| = 18$
   c) $|5x - 20| = 35$ d) $|5x + 20| = 35$

6. Each inequality below has for its solution a set of points on the integer line. Give the midpoint and the two endpoints of each set.
   a) $|x - 2| < 5$ b) $|x - 2| \leq 5$ c) $|x + 4| < 6$
   d) $|x + 4| \leq 6$ e) $|2x - 6| < 12$ f) $|2x + 8| \leq 20$
   g) $|10 - x| < 7$ h) $|3 - x| \leq 9$ i) $|7 + x| < 10$
   j) $|7 + x| \leq 10$ k) $|15 + 3x| < 30$ l) $|21 + 3x| \leq 12$

7. Give the solution set for each equation.
   a) $x + |x| = 0$ b) $x + |x| = 2x$
   c) $x + 3|x| = 4$ d) $x \cdot |x| = -9$
   e) $3x + |x| = 4$ f) $x^2 + |x| = 10$
   g) $2|x| = x + 12$ h) $|x| - 15 = 4x$

2 Construction and Measurement

2–1 Introduction

Our approach to geometry in this chapter will be informal and intuitive. We shall not be concerned with the formulation of precise definitions, or the development of careful proofs. Instead we shall "solve" a number of geometric problems by relying upon our geometric intuition. We shall think of points, lines, and planes as if they were physical objects, points as tiny dots, lines as very narrow streaks, and planes as thin sheets. We assume that students are familiar with terms like

congruent, similar, parallel, perpendicular, symmetric, ...

Although we shall not adopt an axiomatic approach to geometry in this chapter, it is interesting to examine some of the axioms listed by Euclid in the first systematic treatise on geometry, written more than 2300 years ago. The thirteen books of Euclid's *Elements* are a summary of the achievements of Greek mathematicians from the time of Thales (about 600 B.C.) to Euclid (350 B.C.). The first six of these books deal with plane geometry. More than 1000 editions of Euclid's *Elements* have been printed. It is easily the most successful textbook of all time. Literal translations of large sections of the *Elements* occur in many geometry texts used throughout the world.

Euclid listed five *axioms* and five *common notions*. The axioms we wish to consider here are his first three:

  i) To draw a line (segment) from any point to any point, ...

 ii) To extend a line (segment), ...

iii) To draw a circle with any center and any distance, ...

The common notions that we mention are

  i) Things equal to the same thing are equal to each other.

ii) If equals be added to equals the wholes are equal.

iii) If equals be subtracted from equals the remainders are equal.

iv) Things that coincide are equal.

Although Euclid made a distinction between axioms and common notions, suggesting that the common notions are truths of wider applicability than are the axioms, this distinction is not made by modern mathematicians. We shall think of the seven statements above as descriptions of space that we intuitively accept as valid. The first three eventually describe how we can construct geometric configurations. The last four provide the justification for asserting that certain things are true for the figures we draw.

Note that Euclid does not postulate that equilateral triangles or squares can be drawn. In the development of his early theorems he establishes that things like these can be managed by using only the basic construction techniques. When we place compass and straightedge in the hands of students we are giving them the classical Euclidean tools.

An illuminating way to view the geometry of Euclid is to think of a plane as an infinite sheet of paper upon which we have marked two points A and B. Now with our tools, straightedge and compass, we can

(*Axiom* i) Draw the segment from A to B.

(*Axiom* iii) Draw the two circles shown.

(*Axiom* ii) Extend line (segment) AB to cut these circles.

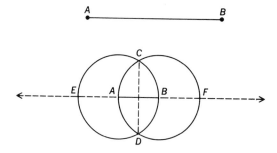

Note that now we have four new points which we have labeled C, D, E, and F. Intuitively, we recognize the following geometric relationships.

1. The segments AC, CB, and BA form an equilateral triangle.

2. The line CD is perpendicular to the line AB.

In fact, in Euclid's Book 1, we find

Proposition 1. *To construct on a given line as base an equilateral triangle.*

Euclid argues:

1. Since C and B are on the circle with center A, AC is equal to AB.

2. Since C and A are on the circle with center B, BC is equal to BA.

3. Since AC and BC are equal to the same thing (namely AB), they are equal to each other.

If one now reapplies Axioms (i), (ii), and (iii) to the set of six points A, B, C, D, E, F above, drawing all segments and circles that these points determine and extending segments to get all possible intersection points, many new points are obtained. Repeating our basic constructions again and again, more and more interesting configurations appear—parallelograms, rectangles, squares, regular hexagons, five-pointed stars, regular polygons of 17 sides, ... Nearly all the geometry of Euclid can be interpreted as a study of the properties of figures that one can obtain by repetition of these basic constructions. Of course, it is not enough to describe a particular construction. One must *prove* (as we illustrated with the equilateral triangle construction) that a particular construction has been achieved.

Exercises 2–1

1. Using your geometric intuition, answer the following questions about the set of 6 points A, B, C, D, E, F determined by drawing the one line and two circles determined by points A and B.

    a) Is $\triangle ECD$ equilateral?

    b) Does line CD bisect $\angle ACB$?

    c) What is the degree measure of $\angle ACB$? $\angle DCB$? $\angle DBF$? $\angle BFD$?

    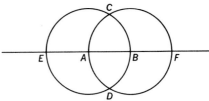

    d) Are lines CA and BD parallel?

    e) Are lines CA and ED perpendicular?

    f) Is quadrilateral $ECFD$ a rhombus?

    g) Are there three points in the figure which determine a right angle?

    h) Is line EC tangent to the circle with center B?

2. Justify your answers to some of the several parts of Exercise 1.

3. Show that the 6 points of the configuration above determine 14 new circles and 9 new lines.

4. With straightedge and compass carefully draw the lines and circles of Exercise 3. How many intersection points do the resulting 16 circles and 10 lines determine?

5. For the configuration of Exercise 4,

   a) Indicate 6 points that form the vertices of a regular hexagon.

   b) Indicate 4 points that are the vertices of a rectangle.

   c) Are there 4 points that are the vertices of a square?

   d) Are there 3 points that determine a 15° angle?

   e) Describe other interesting properties of the set of points in this configuration.

6. Begin with two points and a compass with a fixed setting that enables you to draw the circles determined by these points. Drawing only circles of this size describe the configuration of points that can be obtained in the plane as intersections of these circles. List some interesting properties of this set of points.

When we say that a ruler-compass construction is impossible, we ordinarily mean that beginning with two points and drawing lines and circles as described here one cannot obtain a certain figure. For example, 30° and 15° angles can be obtained, and so 90° and 45° angles can be trisected. Although 60° angles can be drawn, no 20° angles are obtained. A 60° angle cannot be trisected by ruler and compass. Although regular 3-, 5-, and 17-sided polygons occur in our sequence of constructions, no regular 7-, 9-, 11-, or 13-sided polygons put in an appearance.

2–2 Ruler and Compass Constructions

The rules of geometric construction that the Greeks followed may be described by saying that they used an unmarked straightedge and a collapsible compass. The straightedge was not used for measurement, but only to draw lines. A collapsible compass can be used to draw a circle only if the center and a point on the circumference are given; it cannot be used to transfer line segments directly. Euclid's Proposition 2, Book 1, shows that the collapsible compass and straightedge together can be used to transfer segments.

Proposition 2. *To construct at a given point a line equal to a given line.*

This proposition ensures that a modern compass which retains its setting when picked up is no more effective a tool than a collapsible one. The following list of exercises contains constructions with which you are probably familiar. As our discussion of Section 2–1 points out, these constructions could be effected by blindly drawing lines and circles until the desired result is achieved. However, you are expected to exercise some selectivity in deciding what lines and circles to draw.

In the next set of exercises, you are asked to perform several constructions with ruler and compass. We are not concerned with proofs of the validity of the constructions at this point. It is not essential that you be able to do all the constructions at this time. The purpose of the problems is to reestablish your facility to operate quickly and accurately with these tools. This facility will be a great asset in developing geometry informally. You will be able to justify the various constructions later.

Exercises 2–2

1. Perform the construction for the Euclidean proposition stated above. That is, using straightedge and collapsible compass only, construct a line segment with one end point at P whose length is the same as that of AB.

2. Verify the construction of Exercise 1 by giving an argument similar to the Euclidean argument for the construction of an equilateral triangle.

3. Bisect a given segment. $A \bullet\!\!-\!\!-\!\!\bullet B$

4. Construct a perpendicular bisector of a given segment.

5. Construct a perpendicular to a line from a point P not on the line.

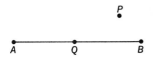

6. Construct a perpendicular to a line at a given point Q on the line.

7. Pretend you have a "rusty" compass that has just one setting and cannot be changed. Use this "rusty" compass (and straightedge) to construct the line that is perpendicular to a line at a given point on the line.

8. Bisect an angle.

9. Copy a given angle.

10. Copy a given triangle.

11. Inscribe a square in a circle.

12. Construct a regular octagon.

13. Construct a regular hexagon.

14. Construct a regular 12-gon.

15. Construct the centroid of a triangle (the point of intersection of the medians).

16. Construct the orthocenter of a triangle (the point of intersection of the altitudes).

17. Circumscribe a circle about a given triangle.

18. Inscribe a circle in a given triangle.

19. Given a circle, construct its center.

20. Construct a 30° angle.

21. Construct a 150° angle.

22. Construct a 75° angle.

23. Construct a parallel to a given line through a given external point.

24. Divide a segment into a given number of congruent segments.

25. Choose any six points on a circle. Name these points, in any order, A, B, C, D, E, and F. Draw the hexagon $ABCDEF$. Consider the intersections of AB and DE, BC and EF, CD and FA. Label them R, S, T, respectively. What is the relationship between the points R, S, T? Try other hexagons with vertices on a circle. What happens if AB and DE are parallel lines? This construction is called *Pascal's magic hexagon*.

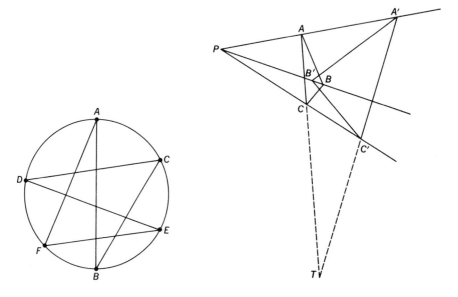

26. Consider $\triangle ABC$ and $\triangle A'B'C'$, where AA', BB', CC' are concurrent lines. If AB and $A'B'$, BC and $B'C'$, CA and $C'A'$ intersect in points R, S, T,

respectively, what is the relationship between R, S, T? Try other such pairs of triangles. What happens if AB and $A'B'$ are parallel? This is called the *Desargues' configuration*.

27. Consider two distinct lines l, l'. Label three distinct points on each line A, B, C on l and A', B', C' on l'. Let R, S, T be the points of intersection of AB' and $A'B$, BC' and $B'C$, CA' and $C'A$, respectively. What is the relationship between R, S, and T? Try other lines and triples of points. This is called a *Pappus configuration*.

2–3 Perimeter and Area

A set of four squares can be arranged to form the two rectangular arrays below.

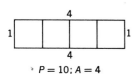

$P = 10; A = 4$

$P = 8; A = 4$

If we choose the edge of one of the four squares as our unit of length, the area of each figure is 4, but in one case the perimeter is 10; and in the other, 8. Other arrangements of the squares are shown below.

$P = 10$

$P = 12$

$P = 16$

As the figures suggest, the maximum perimeter is 16 and the minimum is 8. Show how to arrange the squares and obtain every integral perimeter between 8 and 16.

Exercises 2–3

1. Five unit squares can be arranged to form a region of perimeter 20 or one of perimeter 10. Show how to arrange then so that a perimeter of 13 is obtained.

2. For the accompanying square array, explain how

   a) To remove 4 squares and leave the perimeter the same.

   b) To remove 4 squares and increase the perimeter from 12 to 20.

   c) To rearrange the 9 squares so that the perimeter is 14.

   d) To remove 6 squares and leave the perimeter the same. (In how many ways can this be done?)

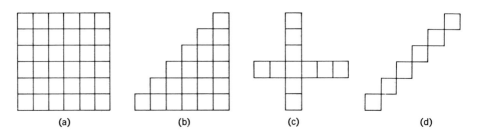

3. Show that all figures below have the same perimeter.

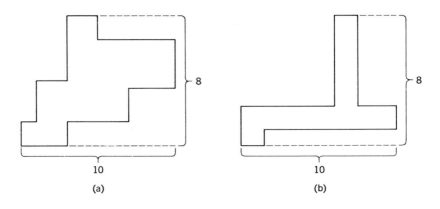

(a)   (b)   (c)   (d)

4. Compute the perimeter of each figure below.

8

10

(a)

8

10

(b)

5. Explain why the perimeter of figure (a) is 2 more than that of figure (b). Give the perimeter of each.

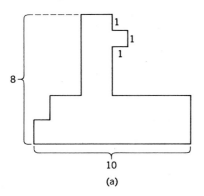

(a)

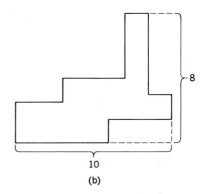

(b)

6. Explain why the perimeter of figure (a) is less than that of figure (b).

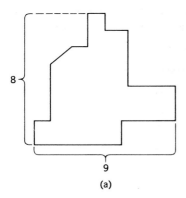

(a)

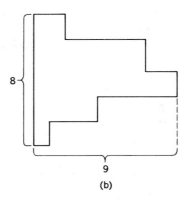

(b)

7. Thirty-six unit squares may be arranged in four different ways to form a rectangle of area 36. Give the perimeter of each rectangle.

8. Show that the parallelogram $ABCD$ and the rectangle $CDEF$ have the same area. Which has the greater perimeter?

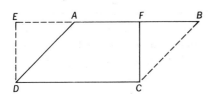

9. Explain why the two following figures have equal areas and equal perimeters. The markings on the sides indicate segments of equal lengths.

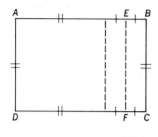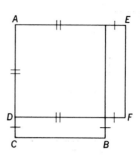

Explain how this diagram shows that if a square has the same perimeter as a rectangle (which is not a square), then the square has a greater area than has the rectangle.

A useful device for illustrating area concepts is a pegboard with pegs arranged as indicated below. Polygonal regions can be delineated by stretching rubber bands or strings around pegs.

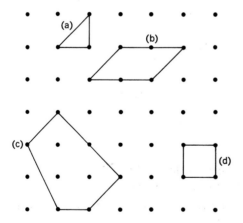

As the figures show, it is easy to compute such areas. If we assign an area of 1 to figure (d), then for (a), (b), and (c) the areas are $\frac{1}{2}$, 2, and 5, respectively. (Verify these results.) To see that the area of figure (a) is $\frac{1}{2}$, we could use paper and cut a square of area 1 along the diagonal and compare the two triangles. In a similar manner, we can show informally that the diagonal of a rectangle divides the rectangle into two triangles of equal area.

Two techniques for computing the area of simple closed polygons formed on the pegboard can be illustrated as follows:

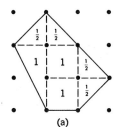
(a)

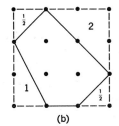
(b)

1. Polygon (a) is dissected by vertical and horizontal lines into unit squares and right triangles. Each right triangle is half a rectangle, and so the area of each part is easily computed. Then the areas of the parts are totalled.

   $1 + 1 + 1 + \frac{1}{2} + \frac{1}{2} + \frac{1}{2} + \frac{1}{2} = 5.$

2. Polygon (b) is inscribed in a square or rectangle.

   Area of the square, 9.

   Area of polygons outside the original polygon, $2 + \frac{1}{2} + \frac{1}{2} + 1 = 4.$

   Area of original polygon, 5.

Using these techniques we can compute the area of any simple closed polygon formed on the pegboard.

With this visual aid many basic ideas stand out clearly. For example, consider the sequence of triangles in the accompanying figure.

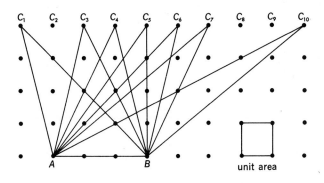
unit area

Show that the area of $\triangle ABC_5$ is 6. Compute the area of each of the other six triangles. What is the area of $\triangle ABC_2$? Of $\triangle ABC_8$? Of $\triangle ABC_9$? What is the relationship between the lines AB and C_1C_{10}? What is the area of $\triangle ABC$ where C is any point on line C_1C_2?

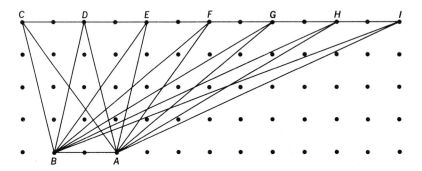

Now consider the sequence of parallelograms $ABCD$, $ABDE$, $ABEF$, $ABFG$, $ABGH$, $ABHI$. Compute the area of each parallelogram. Make a conjecture which is suggested by your calculations.

It is convenient to name the pegs of a pegboard by ordered pairs of whole numbers. For the 10×10 pegboard pictured below we name the pegs as indicated.

Exercises 2–4

1. Compute the area of the triangle with vertices (8, 10), (10, 10), and (10, 7).
2. Compute the area of the triangle with vertices (2, 5), (5, 2), and (3, 9).
3. Note that the line through (3, 9) and (7, 5) is parallel to the line through (2, 5) and (5, 2). It follows that the area of the triangle with vertices (5, 2), (2, 5), and (3, 9) is the same as that of the triangle with vertices (5, 2), (2, 5), and (7, 5). (Why?) Compute the area of this latter triangle.
4. Explain why the area of the triangle with vertices (2, 2), (2, 5), and (7, 5) is the same as that of the triangle with vertices (5, 2), (2, 5), and (7, 5).
5. Show that the triangle with vertices (5, 2), (5, 7), and (10, 7) has the same area as the triangle with vertices (5, 2), (8, 10), and (10, 7). Find two more triangles with the same area as these two.
6. Determine (a, b) so that (a, b), (2, 2), (5, 2), and (7, 5) form a parallelogram. (There are two solutions.)
7. Determine a point (a, b) so that (2, 2), (7, 5), (8, 10), and (a, b) form a parallelogram.
8. Determine (a, b) so that the line through (3, 9) and (a, b) is perpendicular to the line through (5, 2) and (2, 5).
9. Form triangles on a 5 × 5 pegboard (or draw them on squared paper) with areas 1, 2, 3, 4, 5, 6, 7, and 8.
10. Decide for which numbers from 1 through 17 inclusive one can stretch rubber bands about the pegs on a 6 × 6 pegboard and form squares with that area.
11. Find several polygons each of which has 6 pegs on its boundary and 2 in its interior. Show that each has an area of 4.
12. Complete this table determining the areas of several polygons for each of the following cases.

	(a)	(b)	(c)	(d)	(e)	(f)	(g)	(h)	(i)	(j)	(k)
Number of pegs on the boundary	3	4	5	6	3	3	3	4	4	4	12
Number of pegs in the interior	0	0	0	0	1	2	3	1	2	3	4
Area											

13. Guess a formula giving the area of any polygon of the type considered here in terms of the number of pegs that occur on its boundary and in its interior.

14. Explain the derivation suggested below for the area of a triangle in terms of the length b of one side and altitude h to that side.

a) The area of $\triangle ABC$ is half the area of parallelogram $ABCD$.

b) The area of $ABCD$ is the same as the area of rectangle $BEFC$.

c) The area of rectangle $BEFC$ is $b \cdot h$.

d) Hence, the area of $\triangle ABC$ is $\frac{1}{2}bh$.

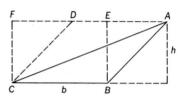

The perimeter of polygon $ABCDEF$ is 7 plus the length of FA. The length of FA is between 2 and 3. (Why?) You may remember that it is the *Pythagorean Theorem* which enables us to compute the precise length of FA. The next set of exercises develops an informal proof of this theorem.

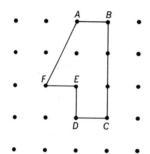

Exercises 2–5

1. Compute the area of each of the four squares in the figure, using AB as the unit. Explain the area formula for squares:

$$A = S^2$$

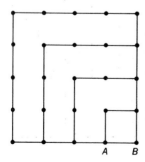

2. Each of the right triangles below has legs of equal lengths. For each compare the area of the square on the hypotenuse with the sum of areas of the squares

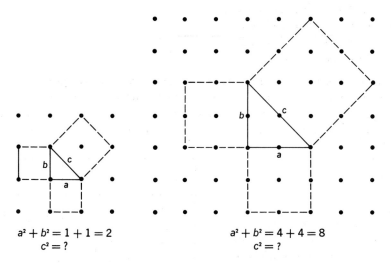

$a^2 + b^2 = 1 + 1 = 2$ $a^2 + b^2 = 4 + 4 = 8$
$c^2 = ?$ $c^2 = ?$

3. Either on a pegboard or on graph paper construct a right triangle with both legs of length 3. Construct the three squares on the legs and hypotenuse, and compare areas as in Exercise 2.

4. The right triangles below have legs of unequal lengths. Compute a^2, b^2, and c^2 in each case, and then compare the area of the square on the hypotenuse with the sum of the areas of the squares on the legs.

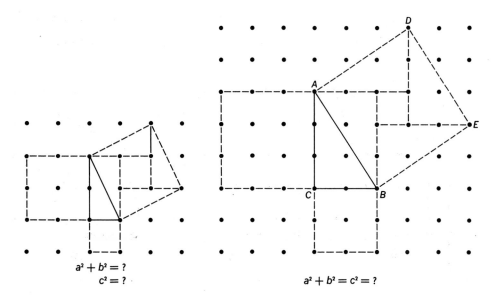

$a^2 + b^2 = ?$ $a^2 + b^2 = c^2 = ?$
$c^2 = ?$

5. Note that in Exercise 4 we can describe the way the sides are drawn for square $ABED$ as follows:

From A to B is *down* 3 and *right* 2,
From B to E is *up* 2 and *right* 3,
From A to D is *up* 2 and *right* 3,
From D to E is *down* 3 and *right* 2.

Either on a pegboard or on graph paper construct two right triangles, one with legs 3 and 4, the other with legs 2 and 5. Construct the squares on the legs and hypotenuse for each triangle. Describe, as above, how the sides are drawn for the squares on each hypoteneuse. Compute a^2, b^2, and c^2 for each triangle.

6. Complete the accompanying table, computing the length of the hypotenuse of each right triangle considered in Exercises 2, 3, 4, and 5.

7. What is the perimeter of polygon $ABCDEF$ in the diagram preceding the section "Exercises 2–5"?

a	b	a^2	b^2	c^2	c
1	1	1	1	2	$\sqrt{2}$
2	2	4	4		
3	3				
1	2				
3	4				
2	5				

8. Find the area and perimeter of each polygon in the accompanying figure, using the indicated unit.

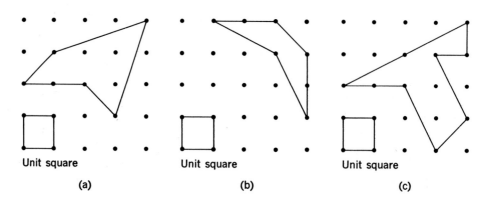

Unit square   Unit square   Unit square

(a)   (b)   (c)

9. A rectangle is 12 in. long and 5 in. wide. Find the dimensions of another rectangle such that the ratio of the perimeter of the second rectangle to the perimeter of the first is 2 to 1, and the ratio of the area of the second to the area of the first is also 2 to 1.

2–4 Volume and Surface Area

A set of four cubes of the same size can be arranged to form the two solids below.

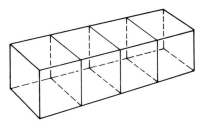

$S = 18$, $V = 4$

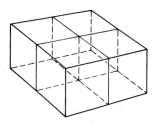

$S = 16$, $V = 4$

If we choose an edge of the cube as our unit of length, then the volume of each solid is 4, but in one case the surface area is 18 and in the other, 16. Consider other arrangements of the cubes. What are the maximum and minimum surface areas? Can you arrange the cubes and obtain every integral surface area between the minimum and the maximum?

Exercises 2–6

1. Show that the minimum and maximum surface areas for figures formed by eight unit cubes are 24 and 48, respectively. Describe arrangements that give areas of 28 and 34.

2. The cube at the right is made up of smaller unit cubes.

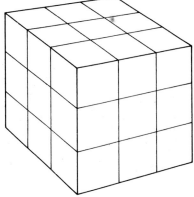

a) Find the surface area and volume of the cube.

b) There are 8 cubes, each of which can be removed without changing the surface area. Explain.

c) There are 12 cubes, each of whose removal increases the surface area by 2 units. Explain.

d) There are 6 cubes, each of whose removal increases the surface area by 4 units. Explain.

e) In parts (b), (c), and (d) we considered 26 of the 27 cubes. What is the situation if the 27th cube is removed?

3. For the cube of Exercise 2, what is the maximum surface area that can be obtained by successively removing cubes? What is the minimum volume that can be attained in obtaining this maximum surface area?

4. What is the surface area of one of the wedges obtained by cutting a unit cube as shown?

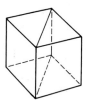

5. Find the volume and surface area of each of the wedges obtained from the rectangular solid.

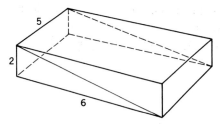

6. Find the volume of each figure. The unit is suggested by the black lines.

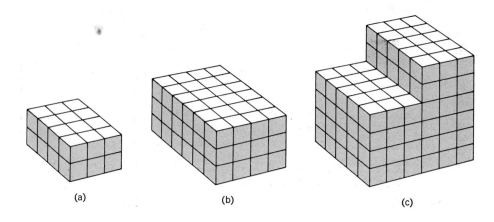

(a)   (b)   (c)

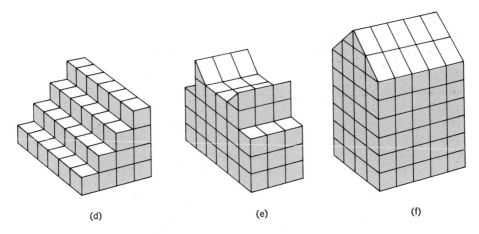

(d)   (e)   (f)

7. Find the volume and surface area of each box. The lengths of the sides are shown.

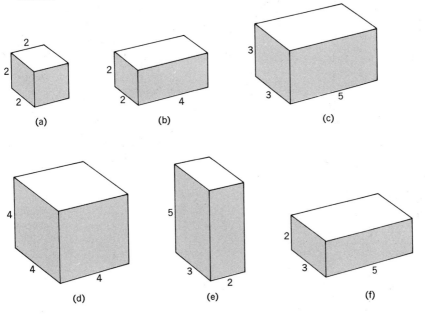

(a)   (b)   (c)

(d)   (e)   (f)

8. a) If each dimension of the boxes in Exercise 7 is doubled, how are the surface areas changed?

   b) If each dimension is tripled, how are the surface areas changed?

   c) If each dimension is doubled, how are the volumes changed?

   d) If each dimension is tripled, how are the volumes changed?

9. Find the volumes.

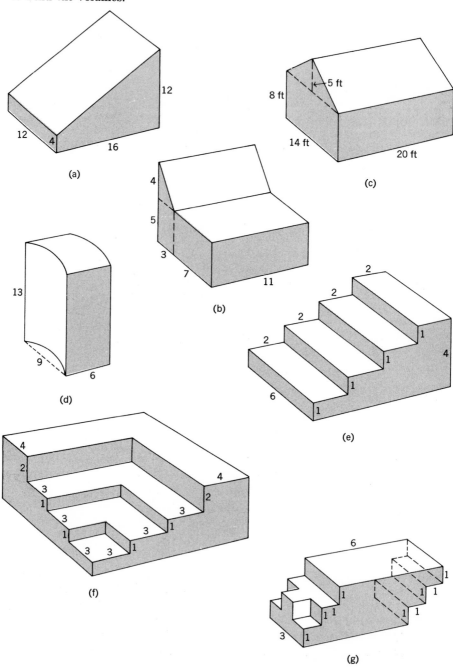

(a)

(b)

(c)

(d)

(e)

(f)

(g)

10. a) What is the volume and surface area of the box of unit cubes?

b) If the cube marked A and the ones below it are removed, what is the volume and surface area of the remaining solid?

11. A straight groove like this is cut into the face of a block 18 in. long. By how much is the surface area of the block increased?

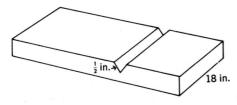

12. What is the volume of Fig. (a) if

a) all cubes are inch cubes?   b) all cubes are 2-in. cubes?

c) all cubes are 3-in. cubes?

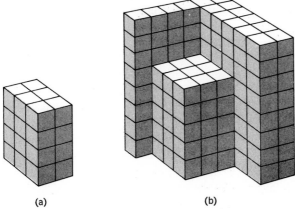

(a)   (b)

13. What is the surface area of Fig. (a) if

a) all cubes are inch cubes?   b) all cubes are 2-in. cubes?

c) all cubes are 3-in. cubes?

14. Repeat Exercise 12, using Fig. (b).

15. Repeat Exercise 13, using Fig. (b).

16. Unit cubes are stacked in the corner of a room as in the picture.

   a) What is the volume?

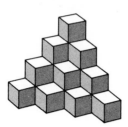

   b) What is the surface area visible in the picture?

17. Consider a cube 3 in. on the edge. If you have a knife and make straight cuts completely through the cube, what is the fewest number of cuts needed to divide the cube into one-inch cubes? How many such cuts are needed to so divide a 10-inch cube?

18. Write a formula for the minimum number of cuts necessary to divide a cube of edge n inches into one-inch cubes.

19. If the cube of Exercise 18 is painted on the outside before it is cut, how many of the n^3 one-inch cubes will be

   a) painted on 3 faces?   b) painted on 2 faces?   c) painted on 1 face?

   d) painted on no face?

2–5 Estimation of Area and Volume

Thus far we have considered area and volume of nicely shaped regions and solids. It is natural to ask for the area and volume of irregular figures.

Estimates of the area of a region can be obtained by placing finer and finer grids of squares over the region. We indicate three stages of this approximation process for a region bounded by an irregular curve. For each grid we count all the squares which contain any point inside the region.

   a) The unit square is $ABCD$. Each of the 16 unit squares contains points in the interior of the region, and the first estimate of the area is 16.

   b) To get a second approximation we count the squares of area $\frac{1}{4}$ (like $EFGH$) which contain interior points of the region. Show that this second estimate is 14.

   c) Count squares of area $\frac{1}{16}$ (like $IJKL$), and compute a third approximation for the area of the region.

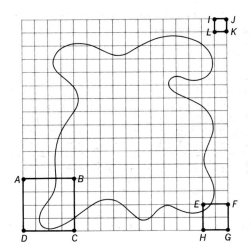

Volumes can be estimated by imagining a region of space successively partitioned into smaller and smaller cubes. This is a theoretically satisfactory method, although of course it has little practical application to the measurement of the volume of actual solids. Explain how volumes of some irregular figures could be approximated by submerging them in water. (Would this method work for a sponge?)

Exercises 2–7

1. For the region of the example imagine the next grid, consisting of squares whose areas are $\frac{1}{64}$ of a unit, and make a fourth area approximation.

2. The area approximations made in the example are obviously all too large. (Explain.) Obtain a sequence of approximations, all of which are too small, by counting only those squares which contain *no points outside the region*. Note that the first estimate is 2.

3. On squared paper draw as large a quarter-circle as possible, choosing a convenient unit of length so that the radius of the circle is 10. According to the area formula for a circle,
$$A = \pi r^2,$$
your quarter-circle should have an area very close to 78.54 square units. (Verify the arithmetic.) Now utilize the squares to make a sequence of 3 or 4 upper approximations to the area of the quadrant and the same number of lower approximations. Take the average of your best upper and best lower approximations and compare with the 78.54 figure. Your figure should differ from this by less than one percent.

4. How could a hollow unit cube, with an open face, that could be filled with water or sand be used to measure the volume of a hollow cone or sphere?

5. Estimates the areas for (a), (b), and (c).

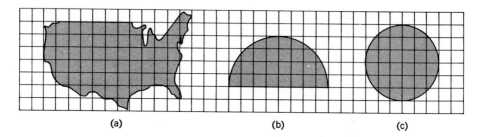

(a)   (b)   (c)

6. Estimate the areas (a) and (b) of the figures on the left below.

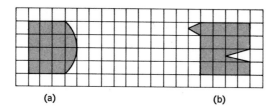
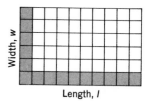

(a)   (b)   Length, *l*

7. Count the squares in the rectangle shown above on the right. How many in a row? How many in a column?
Explain why the area of the rectangle is given by the following formula.
Area $\square = w \cdot l$.

8. Estimate the volume of each solid. Explain your reasoning.

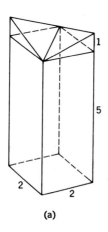
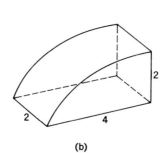

(a)   (b)

9. Estimate the volumes of the following figures.

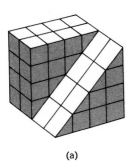

(a)

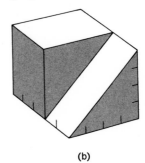

(b)

10. Estimate the volume of each of the solids pictured below.

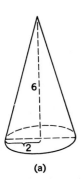

(a)

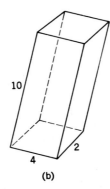

(b)

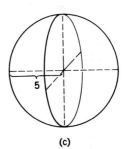

(c)

3 Congruence and Inequalities

3–1 Introduction

Intuitively, two geometric figures are *congruent* if they differ only in their position in space. If an artist made drawings of two congruent figures there would be no way that you could tell which figure a particular drawing represented. Tweedledum and Tweedledee, the Alice in Wonderland characters, were congruent to each other. The artist who illustrated Lewis Carroll's classic never had to worry about which of the twins he was drawing. Two figures which are not congruent may suggest the concept of geometric inequalities. One figure may seem "larger" than the other.

We associate congruence with the concept of *motion*. For example, we can think of triangles as rigid figures formed by fastening thin strips together; then if two triangles are congruent, we can "pick one up" and place it upon the other so that they fit together. Or we can think of the triangles as drawings on separate sheets of transparent paper and visualize one sheet superimposed upon the other so that the drawings "coincide." Recall Euclid's common notion:

Things that coincide are equal.

This physical interpretation of congruence is the one that we adopt in the work of this chapter. We observe that later these notions will need to be refined. When we think of geometric figures as sets of points, then it is difficult to visualize motions. How does one "pick up" the set of points forming one triangle and make it coincide with the set of points forming a second triangle? These are questions for later chapters. For the moment we shall exploit our intuitive notions of congruence. We shall use a notation for lines, segments, rays, and angles that is undoubtedly familiar to you, as shown at the top of page 43. We shall use the symbol ≅ for the *congruence relation*.

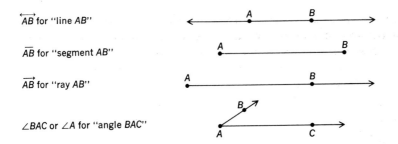

$\overleftrightarrow{AB}$ for "line AB"

$\overline{AB}$ for "segment AB"

$\overrightarrow{AB}$ for "ray AB"

$\angle BAC$ or $\angle A$ for "angle BAC"

3–2 Congruent Segments and Angles

Intuitively, every point is congruent to every other point, every line to every line, and every ray to every ray.

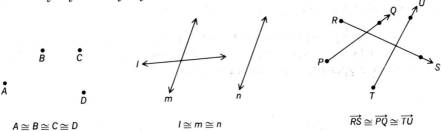

$A \cong B \cong C \cong D$ $\qquad\qquad$ $l \cong m \cong n$ $\qquad\qquad$ $\overrightarrow{RS} \cong \overrightarrow{PQ} \cong \overrightarrow{TU}$

But two segments may or may not be congruent.

$\overline{AB} \cong \overline{CD}$ $\qquad\qquad\qquad$ $\overline{EF} \not\cong \overline{GI};\ EF < GI$

As the diagram above indicates, if two segments are not congruent, one is greater than (>) the other. It is tempting to define congruence in terms of length and say that two segments are congruent if they have the same length. However, in order to describe the measurement process by which one determines the length of a segment, one must use the concept of congruence. For example, if we wish to determine the length of segment $\overline{AB}$ below in terms of unit segment $\overline{XY}$, we consider points $A_1, A_2, A_3, \ldots$ on segment $\overline{AB}$ such that $\overline{XY} \cong \overline{AA_1} \cong \overline{A_1A_2} \cong \overline{A_2A_3} \cong \cdots$

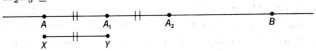

In the figure the *measure* of $\overline{AB}$, denoted by $m(\overline{AB})$, is apparently between 3 and 4. In order to measure the length of a segment we "lay off" congruent copies of the unit segment.

Similarly, when we describe the process by which we compute the measure
of an angle or the area of a region, we find ourselves explaining how congruent
copies of some chosen unit angle or unit region are laid off on the thing being
measured.

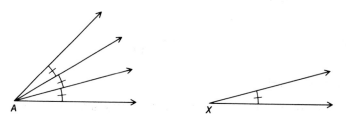

The measure of ∠A in terms of unit ∠X is **3.**

Congruence of segments can be used to define congruence of angles. Intuitively
two angles should be congruent if their rays "open up" in the same fashion. The
diagram below suggests how this intuitive notion can be made precise.

If *A*, *A'*, *C*, *C'*, are chosen so that $\overline{BA} \cong \overline{B'A'}$ and $\overline{BC} \cong \overline{B'C'}$, then there are
three possibilities:

$$\overline{AC} \cong \overline{A'C'}, \qquad \overline{AC} < \overline{A'C'}, \qquad \overline{AC} > \overline{A'C'}.$$

In each of these cases how would you say that the angles compare? Write a
definition for congruent angles in terms of congruent segments.

Aside: The reader who has clear memories of axiomatic approaches to geometry will note the connection between this "definition" of angle congruence
and the basic SAS and SSS theorems for congruence of triangles. In an
elementary geometry course, it is usually assumed that if two triangles have
side, angle, side of one respectively congruent to side, angle, side of the other,
then the third sides are congruent; and if the three sides of one are respectively
congruent to the three sides of the other, then angles opposite congruent
sides are congruent.

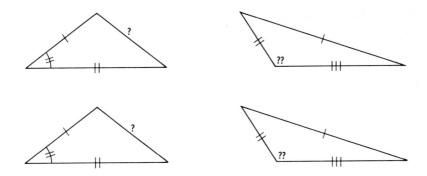

Relate our definition to these SAS and SSS diagrams.

Exercises 3–1

1. Complete each statement below with one of the two statements,

   $\overline{XY} > \overline{AB}$ or $\overline{XY} < \overline{AB}$.

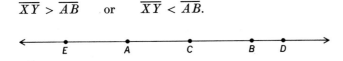

   a) If $\overline{XY} \cong \overline{AC}$, then _____.

   b) If $\overline{XY} \cong \overline{AD}$, then _____.

   c) If $\overline{XY} \cong \overline{AE}$ and $\overline{AE} \cong \overline{AC}$, then _____.

2. Note in Exercise 1 that C is "between" A and B, and B is between A and D. Complete the following statements:

   a) If $\overline{XY} > \overline{AB}$, then there exists a point P on ray $\overrightarrow{AB}$ such that $\overline{XY} \cong \overline{AP}$ and _____.

   b) If $\overline{XY} < \overline{AB}$, then there exists a point P on ray $\overrightarrow{AB}$ such that $\overline{XY} \cong \overline{AP}$ and _____.

3. Four points A, B, C, D are collinear (i.e., lie on one line). It is known that $\overline{AB} < \overline{BC} < \overline{CD}$, $\overline{AD} < \overline{CD}$,   and $\overline{BD} < \overline{CD}$.

   a) Draw a picture showing how this can happen.

   b) Is all the information given above needed in order to determine the picture you drew?

4. Complete each statement below with one of the following statements:

$\angle A \cong \angle A'$, $\angle A < \angle A'$,   or $\angle A > \angle A'$.

a) If $\overline{BC} > \overline{B'C'}$, then _____.

b) If $\overline{BC} < \overline{B'C'}$, then _____.

c) If $\overline{BC} \cong \overline{B'C'}$, then _____.

d) Complete the following statement: $\angle A \cong \angle A'$ if and only if B, C, B', C' can be chosen such that $\overline{AB} \cong \overline{A'B'}$, $\overline{AC} \cong \overline{A'C'}$ and _____.

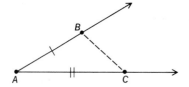
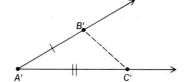

5. Euclid would have considered the figure on the right and concluded that because *the whole is greater than any of its parts*,

$\angle DAC > \angle BAC$.

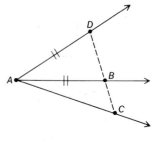

Explain how the reasoning of Exercise 4 enables one to argue formally that $\angle DAC$ is the greater.

6. In the figure for Exercise 5, we say that ray $\overrightarrow{AB}$ is between rays $\overrightarrow{AD}$ and $\overrightarrow{AC}$. Just as we can describe inequalities for segments in terms of betweenness for points, we can describe inequalities for angles in terms of betweenness for rays. Complete the statement below.

$\angle ABC < \angle DEF$ if and only if there exists a ray $\overrightarrow{EG}$ such that

$\angle ABC \cong \angle GEF$ and _____.

Formulate a similar statement beginning, $\angle ABC > \angle DEF$ ....

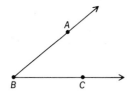
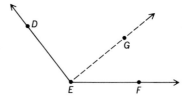

7. If in the diagram $\angle DBA \cong \angle DBC$, what is each of these angles called?

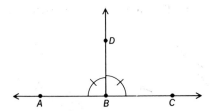

8. Ray $\overrightarrow{BD}$ bisects $\angle ABC$. Explain what this means in terms of congruence.

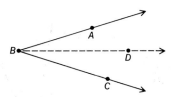

9. Point M is the midpoint of segment $\overline{AB}$. Explain what this means in terms of congruence.

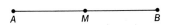

10. Line $\overleftrightarrow{AB}$ is the perpendicular bisector of line segment $\overline{CD}$. Explain what this means in terms of congruence.

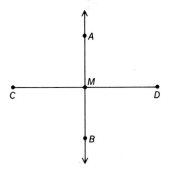

11. Each of the following diagrams suggests an application of Euclid's common notions, (ii) if equals be added to equals ... (iii) if equals be subtracted .... Explain how the common notion is used to draw a conclusion in each case.

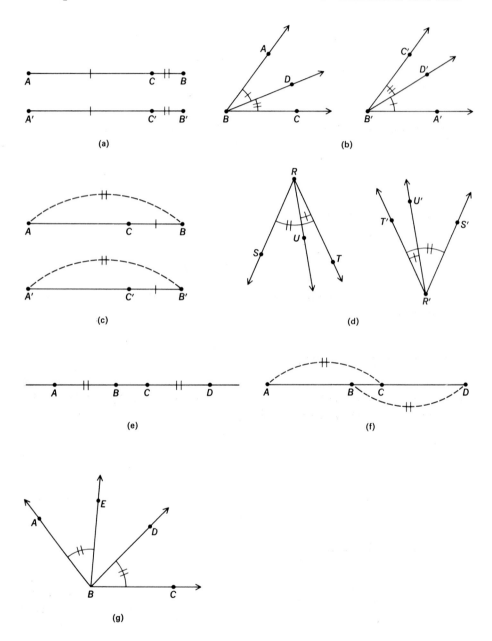

(a)   (b)

(c)   (d)

(e)   (f)

(g)

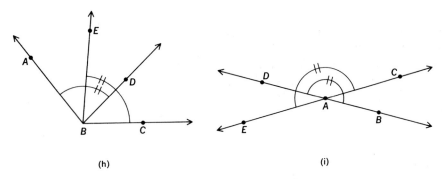

(h)   (i)

12. Consider the set of all points X in the plane such that

$\overline{OX} \cong \overline{OA}$.

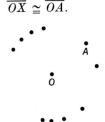

What is this set of points called?

13. Define a square using the concept of congruence.

14. The usual definition of a parallelogram is that it is a *convex* quadrilateral such that opposite sides are parallel. Give an equivalent definition in terms of congruence.

15. If a line is drawn on a sheet of paper and then the paper is folded so that two rays of the line coincide, the crease formed in the paper stands perpendicular to the line. This illustrates Euclid's common notion (iv), *things that coincide are equal.* Explain.

16. If an angle is drawn on a sheet of paper and the paper is folded so that the sides of the angle coincide, the crease formed in the paper bisects the angle. Relate this to Euclid's common notion (iv).

17. Using the definition of congruent angles given earlier, show that the base angles of an isosceles triangle are congruent.

The theorem of Exercise 17 is Euclid's Proposition 5, Book I. This famous theorem received the name *pons asinorum* during the middle ages. The name probably arose from the bridgelike appearance of the figure Euclid used in his proof and from the notion that anyone unable to cross this bridge (i.e., understand

Euclid's proof) must be an ass. The proof
you were encouraged to give in Exercise
17 is intuitively acceptable, but is not
based upon a rigorous axiomatic development. Euclid's proof is quite sophisticated. We shall see more of this theorem
in later chapters.

Euclid's figure for
the *pons asinorum.*

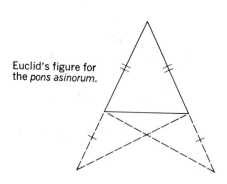

3–3 Congruent Triangles

Two triangles are congruent if their sides and angles can be paired so that corresponding parts are congruent. The notation

$\triangle ABC \cong \triangle DEF$ is an abbreviation for the six statements,

$$\overline{AB} \cong \overline{DE}, \quad \overline{BC} \cong \overline{EF}, \quad \overline{CA} \cong \overline{FD},$$

$$\angle A \cong \angle D, \quad \angle B \cong \angle E, \quad \angle C \cong \angle F.$$

Some of the basic theorems of geometry are those which establish minimum
conditions for the congruence of two triangles. You may remember these theorems:

Side–Angle–Side theorem   (SAS)
Side–Side–Side theorem   (SSS)
Angle–Side–Angle theorem   (ASA)

SAS

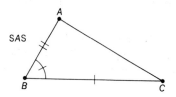 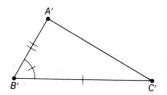

If $\overline{AB} \cong \overline{A'B'}$, $\angle B \cong \angle B'$, and $\overline{BC} \cong \overline{B'C'}$, then $\overline{AC} \cong \overline{A'C'}$, $\angle A \cong \angle A'$,
and $\angle C \cong \angle C'$.

SSS

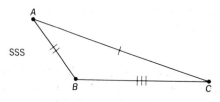 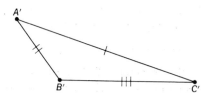

If $\overline{AB} \cong \overline{A'B'}$, $\overline{BC} \cong \overline{B'C'}$, and $\overline{CA} \cong \overline{C'A'}$, then $\angle A \cong \angle A'$, $\angle B \cong \angle B'$,
and $\angle C \cong \angle C'$.

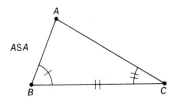

ASA

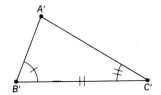

If $\angle B \cong \angle B'$, $\overline{BC} \cong \overline{B'C'}$, $\angle C \cong \angle C'$, then $\overline{AB} \cong \overline{A'B'}$, $\overline{AC} \cong \overline{A'C'}$, and $\angle A \cong \angle A'$.

In the exercises that follow, we can assume that the preceeding three theorems are true.

Exercises 3–2

1. For the triangles on the left below, the statement $\triangle ABC \cong \triangle DEF$ is false. Write a correct statement of congruence.

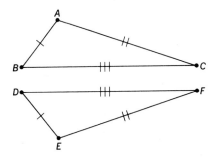
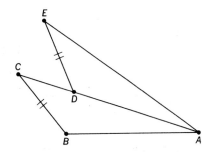

2. Prove that the statement

   $$\triangle CBA \cong \triangle DEA$$

   is false for the triangles at the right above, and write a statement of congruence that might be true.

3. Suppose that for two triangles, $\triangle RST$ and $\triangle UVW$, both the following statements are true:

   $$\triangle RST \cong \triangle UVW, \qquad \triangle SRT \cong \triangle UVW.$$

   What can you deduce? Draw a figure.

4. Suppose that both statements below are true.

   $$\triangle ABC \cong \triangle DEF, \qquad \triangle ABC \cong \triangle EFD.$$

   What can you prove? Draw a figure.

5. For $\triangle ABC$ it happens to be true that $\triangle ABC \cong \triangle BAC$. What can you prove about this triangle? Draw a figure.

6. Suppose that $\triangle DEF \cong \triangle EFD$. What can be proved?

7. Prove that each of the following statements is false for the figure shown:

  a) $\triangle ABC \cong \triangle DBC$ b) $\triangle ABC \cong \triangle CBD$ c) $\triangle ABC \cong \triangle BCD$

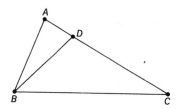

  d) $\triangle ABC \cong \triangle BDC$ e) $\triangle ABC \cong \triangle DCB$ f) $\triangle ABC \cong \triangle CDB$

8. In the rectangular solid shown, explain why

  a) $\triangle EAD \cong \triangle FBC$ b) $\triangle ADF \cong \triangle BCE$

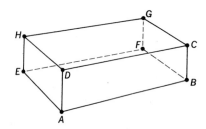

  Name other triangles congruent to $\triangle ADF$.

9. Use six sticks, all the same length, and without breaking any of them place them in such a way that exactly four congruent triangles are formed.

10. Show that if two angles of a triangle are congruent, then the triangle is isosceles.

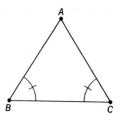

  [*Hint:* Using the figure, show that because of the ASA theorem the statement $\triangle ABC \cong \triangle BAC$ is true.]

  Compare with Exercise 5.

11. Divide the hexagon into (a) two congruent pieces, (b) three congruent pieces, (c) four congruent pieces.

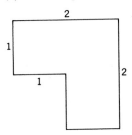

12. Show how to divide the 3 × 8 rectangle into two congruent pieces that can be placed so that they just cover the 2 × 12 rectangle.

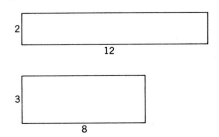

13. Relate the construction of the bisector of an angle to the SSS theorem of triangle congruence.

14. Relate the SSS theorem to the construction of a triangle congruent to a given triangle.

3–4 Right Triangles

We review some useful definitions below.

Two angles with a common side "between" them are *adjacent* to each other:

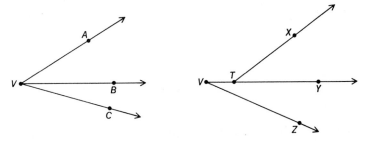

Angle AVB is adjacent to $\angle BVC$. The angles XTY and YVZ are not adjacent. (Why?) Angles AVC and AVB are not adjacent. (Why?)

We say that $\angle AVC$ is a *sum* of the adjacent angles, $\angle AVB$ and $\angle BVC$, and we write:

$$\angle AVB + \angle BVC = \angle AVC.$$

We speak also of sums of nonadjacent angles. If $\angle A$ and $\angle B$ are any two angles, we consider two adjacent angles, $\angle A'$ and $\angle B'$, respectively congruent to $\angle A$ and $\angle B$. The sum of $\angle A'$ and $\angle B'$ is also said to be a sum of $\angle A$ and $\angle B$.

Intersecting lines that form congruent adjacent angles form *right* angles. The sum of two right angles is called a *straight* angle. The sides of a straight angle are *opposite rays*, on a line. Any two angles whose sum is a straight angle are *supplementary* to each other. Two lines which form right angles are *perpendicular* or *orthogonal* ($\perp$) to each other.

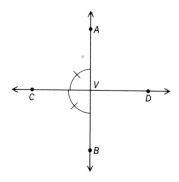

In the figure shown above,

$\angle$'s AVD, AVC, BVC, BVD are *right*,

$\angle$'s AVB and CVD are *straight*,

$\angle$'s AVD and BVD are *adjacent* and *supplementary*,

$\overleftrightarrow{AB} \perp \overleftrightarrow{CD}$.

There is no natural unit segment, but the right angle is a natural unit for measuring angles (Why?) The familiar unit of angle measure, the *degree* (1°) is one-ninetieth of a right angle. Hence the measure of a right angle is 90° and of a straight angle 180°. We write:

$$m\angle RST = 40°$$

to indicate that $\angle RST$ is the sum of 40 angles each congruent to an angle of 1°.

It is interesting to note that Euclid's fourth postulate was

All right angles are equal.

With our present intuitive treatment of geometry this relationship between right angles seems completely obvious.

Angles less than right angles are *acute* angles, and angles greater than right angles and less than straight angles are *obtuse*. If the sum of two acute angles is a right angle the angles are *complementary* to each other.

A triangle that has a right angle is called a *right triangle*. The side of a right triangle opposite the right angle is called the *hypotenuse*, and the remaining sides are called *legs*.

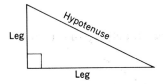

Exercises 3–3

1. In the figure, on the left below, three lines intersect at V.

   a) Name several pairs of adjacent angles.

   b) Name several pairs of nonadjacent angles.

   c) Name several pairs of supplementary angles.

   d) Name several acute angles, obtuse angles.

   e) How many angles are formed by the 6 rays?

   f) Angles AVB and EVD are *vertical* to each other. Name three other pairs of vertical angles.

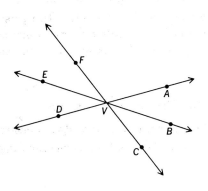   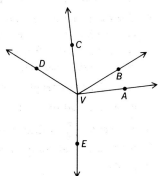

2. Explain how a protractor measures angles by counting the number of one-degree angles that can be laid off on the angle.

3. In the figure on the right above, $\angle AVC$ is a right angle. If the right angle is chosen for the unit, then $m\angle AVC = 1$. Estimate in terms of the right angle unit the measure of

   a) $\angle BVA$,   b) $\angle AVE$,   c) $\angle CVB$,   d) $\angle DVB$.

4. What is true for a pair of angles which are supplements of the same angle? Relate your answer to the theorem:

Vertical angles are congruent.

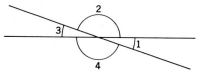

5. Construct (with ruler and compass) right triangles satisfying the following conditions:

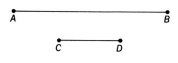

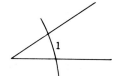

a) One leg congruent to $\overline{AB}$.

b) One leg congruent to $\overline{AB}$ and the other to $\overline{CD}$.

c) Hypotenuse congruent to $\overline{AB}$.

d) Hypotenuse congruent to $\overline{AB}$, and one leg congruent to $\overline{CD}$.

e) One leg congruent to $\overline{AB}$, and one angle congruent to $\angle 1$.

f) Hypotenuse congruent to $\overline{AB}$, and one angle congruent to $\angle 1$.

6. For each of the six constructions of Exercise 5, decide whether two persons performing the construction separately would necessarily construct congruent right triangles. State three sets of *minimal* conditions for the congruence of two right triangles.

7. Describe the triangles formed by

a) a diagonal of a square,   and   b) a diagonal of a rectangle.

8. Segment $\overline{AB}$ is drawn on a sheet of paper with B at the *edge*. Construct with ruler and compass a right triangle with $\overline{AB}$ as a leg and the right angle at B.

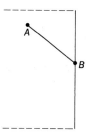

9. Show how to form a square by fitting together 20 right triangles congruent to the one shown below.

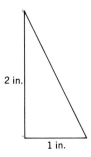

2 in.

1 in.

10. Draw a triangle, and using a protractor draw the rays which trisect each angle. Consider the triangle determined by XYZ, the points of intersection of "adjacent" trisectors. Make a conjecture. (This is known as *Morley's theorem*.)

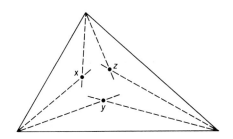

3–5 The Exterior Angle Theorem

Consider $\triangle ABC$. Extend $\overline{AB}$ through B, and let Y be any point on the extension. The angle $\angle CBY$ is called an *exterior* angle of $\triangle ABC$. Similarly, $\angle ZAB$ is an exterior angle of the triangle. By extending $\overline{CB}$ beyond B, a second exterior

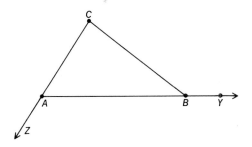

angle with vertex at B is formed. How many exterior angles has a triangle? How do two exterior angles at one vertex compare?

The angles of the triangle at A and C are called *remote interior* (or *opposite interior*) angles with respect to an exterior angle at B. What angles are remote interior angles with respect to $\angle ZBA$?

Draw several triangles of different shapes and compare several exterior angles with their remote interior angles. Your experimental results should agree with the *exterior angle theorem.*

Any exterior angle of a triangle is greater than either remote interior angle.

Hence, in the figure above we have

$$\angle CBY > \angle A, \quad \text{and} \quad \angle CBY > \angle C.$$

What are the corresponding inequalities involving $\angle ZAB$?

In the next set of exercises we shall assume the truth of this theorem and indicate how it may be used to establish many geometric results.

Exercises 3–4

1. Why is it the case that if a triangle has one right angle its other angles are acute?

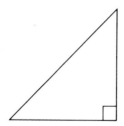

2. Prove that the base angles of an isosceles triangle are acute. [*Hint:* By Exercise 1 they cannot be right. Why cannot they be obtuse?]

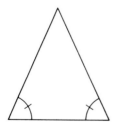

3. Explain why the angles marked at C and D cannot be congruent.

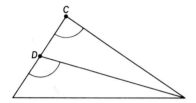

4. Which is greater, $\angle G$ or $\angle CAK$? Give a proof. [*Hint:* Compare $\angle G$ with $\angle CBA$.]

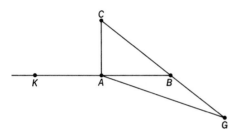

5. Is the following statement true? "Any exterior angle of a quadrilateral is greater than each of the three remote interior angles."

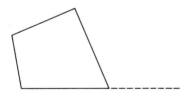

6. In the figure $\overline{AB} \cong \overline{A'B'}$ and $\overline{CD} \cong \overline{B'C'}$. We call $\overline{A'C'}$ a *sum* of $\overline{AB}$ and $\overline{CD}$ and write

$$\overline{AB} + \overline{CD} = \overline{A'C'}.$$

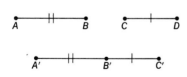

What does it mean to say that the sum of two segments is greater than a segment $\overline{XY}$?

7. Draw several triangles and, using a compass, compare the *sum* of the two shorter sides with the third. Explain the theorem known as the *triangle inequality:*

   Any two sides of a triangle taken together are greater than the third.

   Relate this theorem to the assertion, "A straight line is the shortest distance between two points."

8. Let P be any point in the interior of $\triangle ABC$. Draw $\overline{AP}$ and $\overline{BP}$.

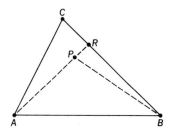

   a) Compare $\angle C$ and $\angle APB$.

   b) Compare $\overline{AC} + \overline{CB}$ and $\overline{AP} + \overline{PB}$.

   c) Prove that $\angle C < \angle APB$ by comparing each angle with $\angle PRB$.

   d) Prove that $\overline{AC} + \overline{CB} > \overline{AP} + \overline{PB}$ by comparing each sum with $\overline{AR} + \overline{RB}$.

9. Use the exterior angle theorem and the figure to show that the angle opposite the greater side of a triangle is the greater angle (i.e., show $\angle ACB > \angle B$).

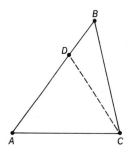

   $$\overline{AB} > \overline{AC}, \quad \overline{AD} \cong \overline{AC}.$$

10. State the converse of the theorem of Exercise 9. Is this theorem true? Can you prove your conjecture?

4 Parallel and Perpendicular Lines and Planes

4–1 Introduction

In this chapter we consider some special properties of parallel and perpendicular lines and planes. The famous Euclidean parallel postulate is discussed, and an approach to parallel lines is presented. Conditions for parallelism and perpendicularity are developed intuitively in terms of *slope*, using the pegboard and graph paper. Properties of parallelograms are investigated. Finally, the angle-sum theorem for triangles is derived and the dependence of this theorem upon the parallel postulate is discussed.

4–2 Slope of a Line

Represent a line l by stretching a string on a pegboard, or picture l on graph paper. A *legal move* on l consists of beginning at a point on l, moving vertically (up or down), and then moving to the right ending on l. In the diagram every legal move must begin with a downward move. (Why?) For example, in the figure a move from A to D and then from D to B constitutes a legal move. We describe this move as "down 1, right 2."

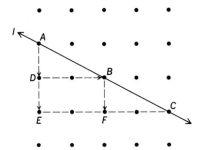

Note that down 2, right 4 is also a legal move. Name other legal moves. If you begin at any point on l, will down 1, over 2 be a legal move on l? Will down $\frac{1}{2}$, over 1 be a legal move? Note that for all legal moves we have used a ratio of vertical distance to horizontal distance of $\frac{1}{2}$.

$$\frac{1}{2} = \frac{2}{4} = \frac{\frac{1}{2}}{1} = \cdots.$$

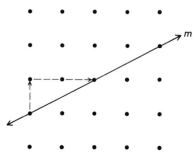

For the line m pictured on the right, legal moves are

up 1, right 2,   up 2, right 4 ....

For m, the ratio of vertical distance to horizontal distance is also one-half. We distinguish between the lines l and m by using both positive and negative numbers. If the vertical movement is upward, we describe it by a positive number; if downward, by a negative number. The resulting (signed) ratios of vertical to horizontal movement are called *slopes* of the associated lines.

$$\text{Slope of a line} = \frac{\text{length of vertical move } (+ \text{ or } -)}{\text{length of horizontal move (to the right)}}.$$

Exercises 4–1

1. Find the slope of the lines containing the indicated segments.

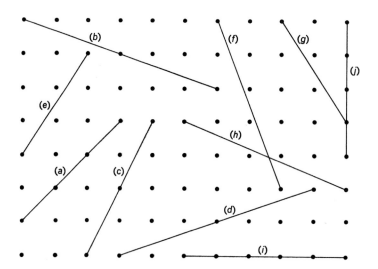

2. For line (i) a legal move can be described as up 0, right 1; up 0, right 2; up 0, right 10; .... What is the slope of every horizontal line?

3. For line (j) a legal move can be described as up 2, right 0; down 4, right 0; .... In these cases the slope ratios cannot be computed. (Why?) What can you say about the slopes of lines that are almost vertical? We do not assign a slope to vertical lines.

4. Only horizontal moves to the right were permitted in legal moves previously. Explain how left horizontal moves could be incorporated into legal moves so that the slopes of the lines would be unchanged.

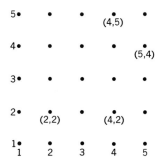

5. Using ordered pairs of whole numbers to name pegs as shown in the figure above, find the slopes of lines touching the following pegs:

   a) $(2, 2)(5, 4)$ b) $(4, 5)(1, 2)$

   c) $(3, 1)(1, 4)$ d) $(2, 2)(5, 2)$

   e) $(4, 1)(4, 4)$ f) $(5, 3)(1, 5)$

6. On graph paper label points with ordered pairs of whole numbers and draw the lines with the indicated slopes through the given points.

   a) $(1, 5)$ slope $-2/3$ b) $(2, 1)$ slope $1/4$ c) $(3, 1)$ slope -2

   d) $(1, 1)$ slope 4 e) $(3, 3)$ slope 0 f) $(5, 1)$ slope undefined

4–3 Perpendicular Lines

If we look at lines on the pegboard or on graph paper, we can get intuitively simple conditions for perpendicularity in terms of slope. Let us take two lines, one vertical and the other horizontal, which are perpendicular. Note that if you look at such a pair of lines and mentally label one as l and the other as m, and then give the paper

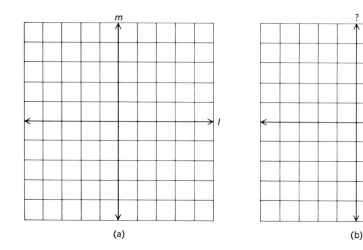

(a)   (b)

a quarter-turn about the point of intersection, the two lines exchange position. In (b) how do you decide which line is *l* if you do not know whether or not the paper has been given a quarter-turn? Explain how this illustrates that for lines *l* and *m* adjacent angles are congruent.

When we look at any two intersecting lines on graph paper they will be perpendicular if and only if the lines exchange position when the paper is given a quarter turn about the point of intersection. (Why?) We can use this concept to derive the relationship between the slopes of two perpendicular lines.

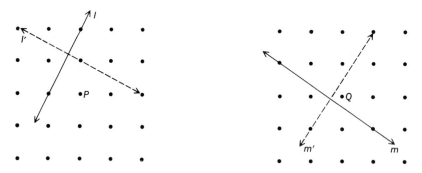

Legal moves on *l* and *l′* are up 2, right 1 and down 1, right 2, respectively. What are legal moves on the lines *m* and *m′*. Show that a 90° rotation about *P* takes *l* to *l′*. How is *m′* related to *m*?

The following figures apply to the exercises in the next section. These exercises develop ideas of slope related to perpendicular lines. Be sure to work both Exercise 1 and Exercise 2.

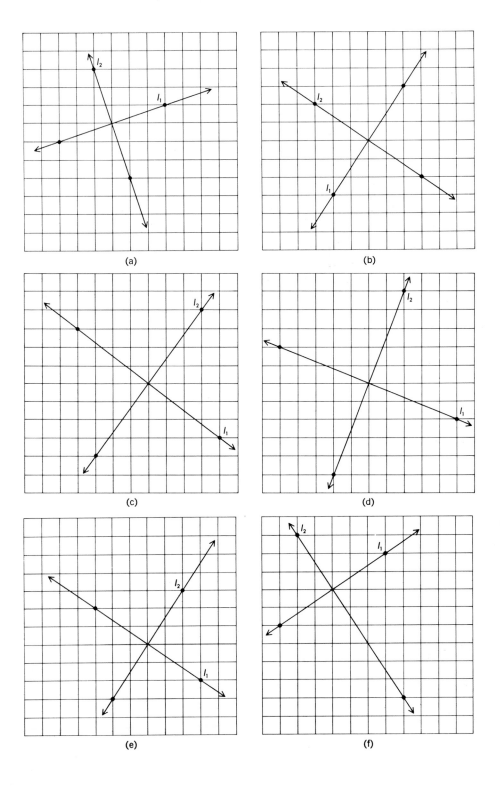

(a)

(b)

(c)

(d)

(e)

(f)

Exercises 4–2

1. a) In each case $l_1 \perp l_2$. Fill in the chart below.

	(a)	(b)	(c)	(d)	(e)	(f)
Slope of l_1						
Slope of l_2						

b) For the six lines of Exercise 1, turn your book so that the horizontal direction becomes vertical. With this change in your frame of reference, complete a table as in 1(a).

2. On a pegboard or on graph paper construct other pairs of perpendicular lines. How are the slopes of perpendicular lines related?

a) Given that $l_1 \perp l_2$. Complete the following chart.

Slope of l_1	$\frac{5}{6}$	$-\frac{4}{3}$			$\frac{9}{7}$		$-\frac{18}{19}$
Slope of l_2			5	$-\frac{1}{7}$		40	

b) If $l_1 \perp l_2$, neither line is vertical, and their slopes are respectively m_1 and m_2, what is $m_1 \cdot m_2$?

3. Let the ordered pairs of whole numbers represent pegs on a peg board or lattice points on graph paper. Show that the line on the first two points is perpendicular to the line on the second and third.

   a) (2, 17), (5, 16), (3, 10)   b) (9, 8), (13, 11), (10, 15)

   c) (6, 6), (12, 9), (15, 3)   d) (1, 6), (4, 3), (6, 5)

4. Can you find a technique for computing the slope of a line directly from the ordered pairs of numbers representing the pegs?

5. Use the diagram at the right to argue that it is impossible to draw two perpendiculars to a line l from a point P not on l. [*Hint:* Remember the exterior angle theorem.]

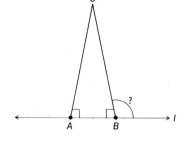

6. If the length of segment $\overline{PQ_3}$ is the shortest distance from P to line l, what can you say about lines $\overleftrightarrow{PQ_3}$ and l?

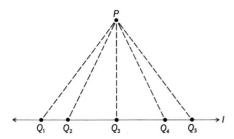

7. Draw a segment $\overline{AB}$ and construct its perpendicular bisector l.

   a) Let P be any point of l. Compare $\overline{PA}$ and $\overline{PB}$.

   b) Make a conjecture.

   c) State the *converse* of your conjecture. Is this converse true?

   d) What have the SAS and SSS theorems to do with parts (b) and (c)?

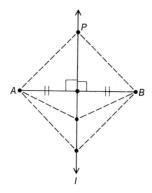

8. An altitude of a triangle is a line on a vertex perpendicular to the opposite side. Construct the three altitudes of a triangle. (This is conveniently done on graph paper.) Make a conjecture.

9. The point of *concurrency* of the altitudes of a triangle is called the *orthocenter* of the triangle. What is true for the orthocenter of an obtuse triangle? A right triangle?

10. Construct the three perpendicular bisectors of the sides of a triangle. What appears to be true for these lines?

11. The point of concurrency of the perpendicular bisectors of the sides of a triangle is called the *circumcenter* of the triangle. Explain the choice of this name.

12. In what kind of triangle is the orthocenter a vertex of the triangle? In what kind of triangle do the orthocenter and circumcenter coincide?

4-4 Perpendicular Planes

Let a sheet of paper represent a plane. Draw at least three concurrent lines on the paper. Place the tip of your pencil at the point of intersection of these lines so that the pencil is perpendicular to one of the lines. Is it necessarily perpendicular to the others? Now hold the pencil so that it stands perpendicular to two of the lines. Is it necessarily perpendicular to the others?

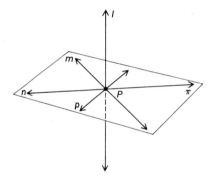

We say that a line l is perpendicular to a plane π if l is perpendicular to every line in π which is *incident* with the point of intersection of l and π. In the figure, l intersects π at P and apparently is perpendicular to every line of π through P. It is intuitively clear that a "vertical" line is perpendicular to a "horizontal" plane. If a line is perpendicular to a plane, we say also that the plane is perpendicular to the line and write $l \perp \pi$ or $\pi \perp l$.

Your experiment of the first part of this section suggests that if a line is perpendicular to two intersecting lines in a plane (at the point of intersection), then the line is perpendicular to the plane.

At this point it is natural to ask what is meant by perpendicular planes. A glance at the relation between a side wall of a room and the floor or ceiling indicates how perpendicular planes should look. It would be convenient to have a definition of perpendicular planes analogous to that given for perpendicular lines. However, before making such a definition we must discuss the *angles* formed by planes.

Just as two intersecting lines form 4 angles (which are not straight angles), two intersecting planes form 4 *dihedral angles* (which are not *flat* dihedral angles). Each of these dihedral angles consists of two *half-planes* and their common edge.

In the figure we could speak of dihedral angle *D-AB-E*, or *D-AB-F*. Each half-plane is called a *face* of its dihedral angle. A *plane angle* of a dihedral angle is an angle formed by two rays perpendicular to the edge at the same point, with one ray in each face. In the figure, ∠*DCE* is a plane angle of dihedral angle *D-AB-E*, but ∠*DCF* is not a plane angle of dihedral angle *D-AB-F*. (Why?) What seems to be true for any two plane angles of a dihedral angle?

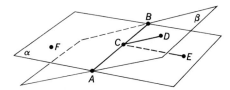

We agree that the *measure* of a dihedral angle is the measure of any of its plane angles. It is clear now how *right* dihedral angles and *perpendicular planes* are defined. Formulate these definitions.

Exercises 4–3

1. In the figure a cube rests on a plane π. Line $\overleftrightarrow{CB}$ is perpendicular to the plane.

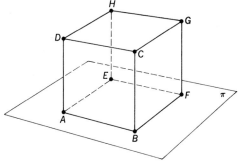

a) What other lines shown are perpendicular to π?

b) The plane of face *ABCD* is perpendicular to π. Name three other planes perpendicular to π.

c) How are lines $\overleftrightarrow{AE}$ and $\overleftrightarrow{AD}$ related? Lines $\overleftrightarrow{AB}$ and $\overleftrightarrow{AD}$?

d) Name two planes perpendicular to face *CDHG*.

e) Name two right dihedral angles and a plane angle of each.

2. Use a pencil or ruler to represent a line perpendicular to a table top. Use a piece of carboard to represent a plane that contains this line. Why is the cardboard perpendicular to the table top? Use a second pencil to form the plane angle of a dihedral angle formed by the planes. Is this a right angle? Explain.

3. Planes α and β are perpendicular, with $\angle ABC$ a plane angle of dihedral angle $C\text{-}GB\text{-}A$. Complete each statement below by supplying the correct symbol, $<$ or $\cong$ or $>$, for each blank.

a) $\angle ABE$ _____ $\angle ABC$.   b) $\angle FBE$ _____ $\angle ABC$.

c) $\angle DBE$ _____ $\angle ABC$.   d) $\angle DBC$ _____ $\angle ABC$.

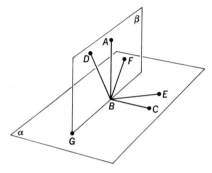

4. a) At a given point on a plane how many lines are perpendicular to the plane?

b) How many planes are perpendicular to a line at a given point on the line?

c) Line l is in plane π. How many planes contain l and are perpendicular to π?

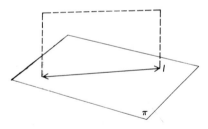

5. Planes α and β intersect in l. Line m in β is perpendicular to l. Is β necessarily perpendicular to α? Explain.

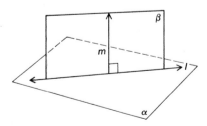

6. Line l intersects plane π at P. Is there a plane containing l and perpendicular to π? Explain.

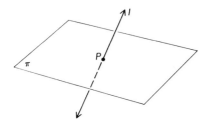

7. In a rectangular solid, adjacent faces are perpendicular. $\overline{DG}$ is a diagonal of the solid, and $\overline{EA}$ is a diagonal of the face $ADEF$. $\overline{EA}$ is not a diagonal of the solid.

   a) How many diagonals has the solid?

   b) Are these diagonals concurrent?

   c) The diagonals lie in pairs in 6 planes. Describe these planes.

   d) Each diagonal of the solid is the hypotenuse of 6 right triangles whose legs are respectively a diagonal of a face and an edge. Name these right triangles for diagonal $\overline{DG}$.

   e) If the dimensions of the solid are 2, 3, and 6, what is the length of $\overline{DG}$?

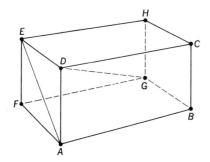

8. Describe minimal conditions for determining a unique plane. (In other words, how many points and lines are necessary?)

4–5 Parallel Lines and Planes

The notion of parallel lines is a familiar one. Give several physical examples that suggest parallel lines. Euclid defined parallel lines as *coplanar* lines (lines in the same plane) which do not intersect. Two lines in space which are not coplanar are called *skew* lines. (Could skew lines intersect?) Under what condition is a line parallel to a plane? When are two planes parallel? We use the symbols ∥ and ∦ as abbreviations for "is parallel to" and "is not parallel to."

Exercises 4–4

1. In the rectangular solid,
    a) Name all pairs of parallel lines.
    b) Name several pairs of perpendicular lines.
    c) Name several pairs of skew lines.
    d) Name several pairs consisting of a plane and a line parallel to the plane.

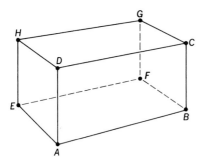

2. Represent several pairs of parallel lines on a pegboard or on graph paper. What condition must the slopes of two lines satisfy in order that the lines be parallel?

3. The ordered pairs of whole numbers can be interpreted as pegs on a pegboard or as points on graph paper. Determine whether or not the line on the first two points is parallel to the line on the second pair of points.
    a) (2, 3) and (8, 9), (14, 3) and (18, 7).
    b) (3, 10) and (5, 4), (11, 10) and (13, 4).
    c) (7, 2) and (15, 8), (3, 1) and (9, 9).

4. If $l \nparallel m$ and lines l and m are coplanar, what can you assert for these lines?

5. If $l \nparallel m$ and l and m do not intersect, what can you say about them?

6. If $m \parallel n$ and $l \perp m$ and the lines are coplanar (see figure at the top of page 73), what can you say about l and n?

7. Suppose that $m \parallel n$ and $n \parallel l$.
    a) What can you say about m and l if the three lines are coplanar?
    b) What if the three lines are not coplanar?

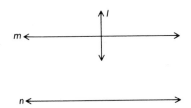

8. In the figure, line t' is a *transversal* of lines l and m, while t is not. Write a definition of a transversal of two lines.

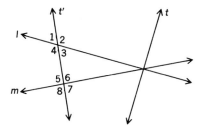

9. When two lines are cut by a transversal, the following terminology is used to describe the eight angles that are formed:

exterior angles, *interior* angles, *corresponding* angles, *alternate interior* angles, *alternate exterior* angles.

In the figure for Exercise 8,

a) Name the four exterior angles.

b) Name the four interior angles.

c) Angles 1 and 5 are "correspondingly placed" and are said to be a pair of corresponding angles. Name three other such pairs.

d) Angles 4 and 6 are a pair of alternate interior angles. Name a second pair.

e) Name two pairs of alternate exterior angles.

10. A transversal t of lines l and m forms congruent corresponding angles as shown. Use the exterior angle theorem to prove that lines l and m cannot intersect on the "right" of the transversal t. Why cannot they intersect on the left?

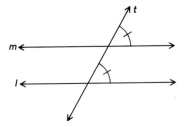

11. Let m and n be two lines in space. From points A and C on m drop perpendiculars to n at B and D, respectively.

a) If $\overline{AB} > \overline{CD}$ and m, n are coplanar, what can you conclude?

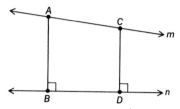

b) If $\overline{AB} > \overline{CD}$ and m, n do not intersect, what can you conclude?

c) Is it possible that m and n are skew and $\overline{AB} \cong \overline{CD}$?

d) Is it possible that $m \parallel n$ and $\overline{AB} \not\cong \overline{CD}$?

12. If two parallel planes are cut by a third plane, what can you say about the lines of intersection? Prove your assertion.

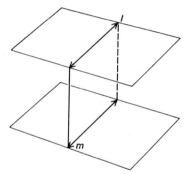

13. Lines m and n intersect, as do m' and n'. Also, $m \parallel m'$ and $n \parallel n'$. The four lines are not coplanar. What can you say about the two planes determined by these pairs of intersecting lines?

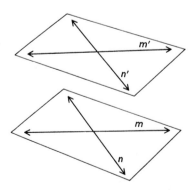

14. Planes α and β (shown below on the left) are parallel, and line $\overset{\leftrightarrow}{AB} \perp \alpha$. Explain why $\overset{\leftrightarrow}{AB} \perp \beta$. [*Hint:* Consider two planes each containing $\overset{\leftrightarrow}{AB}$.]

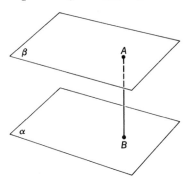 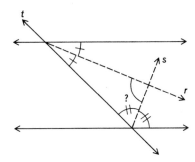

15. Line t (shown above on the right) is a transversal of two parallel lines. Rays r and s bisect the indicated interior angles. What can you say about the angles formed by these bisecting rays?

16. Hold two pencils or yardsticks in a position to represent skew lines. Try to visualize a third line perpendicular to both. Does every pair of skew lines have a common perpendicular? Could a pair of skew lines have more than one common perpendicular?

17. Describe "all ways" that three lines can be related in space in terms of whether or not they intersect or are coplanar.

4–6 The "Parallel" Postulate

In our intuitive work with pegboards and graph paper we have assumed properties of parallel lines that are not justified by those axioms of Euclid which we listed earlier. In particular, when we work with graph paper and ideas of slope, we accept as a fact that if P is not on a line l, then there is just one line through P parallel to l. It is "intuitively clear" that there cannot be two *different* horizontal lines through P. However, if we turn from physical representations of lines and look at the logical structure of geometry as established by the axioms of Euclid,

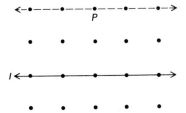

we find that in order to establish many of the results concerning parallel lines that we have intuitively accepted in this chapter, another assumption is needed. We have already mentioned four of Euclid's postulates. The famous *fifth postulate* of Euclid was the center of a mathematical controversy that raged for many years. Mathematicians were not able to settle their differences of opinion concerning this postulate until the 19th century.

We have already seen that if a transversal of two lines forms a congruent pair of corresponding angles, the lines are parallel. An equivalent assertion is that if the interior angles, $\angle 2$ and $\angle 3$, are supplementary, then the lines are parallel. (Why are these two assertions equivalent?)

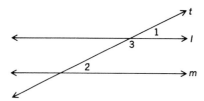

If $\angle 1 \cong \angle 2$, then $l \parallel m$. If $\angle 2 + \angle 3$ is
a straight angle, then $l \parallel m$.

Now we turn to the converse statement. Suppose that in the figure above, $l \parallel m$. Then can we be sure that $\angle 2$ and $\angle 3$ are supplementary? This is essentially Euclid's fifth postulate:

If a line cuts two lines and makes the sum of the interior angles on one side less than two right angles, then the lines will intersect on that side.

A version of Euclid's postulate that may be familiar to you is *Playfair's axiom*:

If P is not on a line l, then there is only one line on P parallel to l.

For 2000 years from the time that Euclid stated his fifth postulate many mathematicians attempted to base a proof of Euclid's parallel postulate upon his other assumptions. All attempts failed. Finally it was shown in the 19th century by the mathematicians Gauss, Bolyai, Lobachevski, and others that we can accept all Euclidean assumptions except the fifth postulate, substitute for the fifth postulate an assumption that contradicts it, and obtain a non-Euclidean geometry which differs from Euclidean geometry. This new geometry is logically as sound as Euclidean geometry and quite useful. For example, if we assume that there is more than one parallel to a line through a point not on the line the geometry which results is called *Lobachevskian* (or hyperbolic). This geometry was developed by Gauss in his private researches and was described by N. Lobachevski in his published work *Theory of Parallels* (1834). This is a geometry that has proved useful in analyzing Einstein's theory of relativity in physics.

In the next group of exercises, four theorems are stated. Each is a consequence of Euclid's parallel postulate. Moreover if we accept any of these theorems as true, we can deduce Euclid's postulate. We can say that each of these four statements is *equivalent* to Euclid's fifth postulate. Choose one of the four exercises and (a) use Euclid's fifth postulate to prove the theorem, and then (b) accept the theorem as true and use it to establish Euclid's postulate.

Exercises 4–5

1. If two parallel lines are cut by a transversal t, then any two interior angles on the same side of t are supplementary.

2. If two parallel lines are cut by a transversal, then any two alternate interior angles are congruent.

3. In a plane if a line intersects one of two parallel lines, then it intersects the other.

4. In a plane if a line is perpendicular to one of two parallel lines, then it is perpendicular to the other.

4–7 Transversals of Parallel Lines

In this section we gather together a few of the results for parallel lines that follow from Euclid's fifth postulate. We need these results for the study of parallelograms. These results are suggested in the exercises below.

Exercises 4–6

1. Construct two parallel lines labeling angles as in the figure.

   a) Measure and compare $\angle 3$ and $\angle 4$. What does this result illustrate?

   b) Measure and compare $\angle 1$ and $\angle 2$. What does this illustrate?

   c) Compute the sum $\angle 1 + \angle 4$. What does this illustrate?

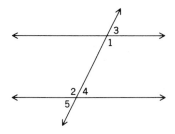

2. If $l \parallel m$ and $p \parallel n$,

   a) Show that all odd-numbered angles are congruent.

   b) How are any two odd- and even-numbered angles related?

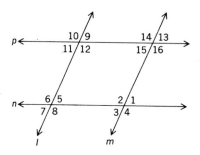

3. Construct a pair of angles with the sides of one parallel to the sides of the other. Compare the angles and make a conjecture.

4. Relate Exercises 3 and 2 by noting that the sides of $\angle 10$ in Exercise 2 are parallel to the sides of $\angle 1$. (Explain.) Do you need to revise your conjecture of Exercise 3?

4–8 Parallelograms

A quadrilateral is a four-sided polygon. In the figures, $ABCD$ and $EFGH$ are quadrilaterals, the first is *convex* and the second, *nonconvex*. Define a convex quadrilateral. Name two *opposite* sides of each quadrilateral; two *opposite* angles; two *adjacent* (consecutive) sides; two *adjacent* (consecutive) angles. Segments $\overline{AC}$ and $\overline{FH}$ are diagonals. Name the other diagonals.

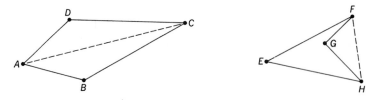

   In this section we consider quadrilaterals with parallel opposite sides. A *trapezoid* is a quadrilateral with exactly one pair of opposite sides parallel. A *parallelogram* is a quadrilateral with both pairs of opposite sides parallel. A *rhombus*, *rectangle*, and *square* are special parallelograms. Define each of these last three figures.

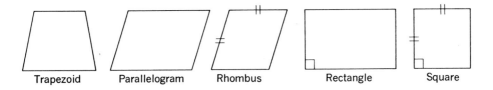

Trapezoid   Parallelogram   Rhombus   Rectangle   Square

Exercises 4–7

1. The ordered pairs of whole numbers represent pegs on a pegboard or lattice points on graph paper. Consider each set of four as consecutive vertices of a quadrilateral, find the slopes of opposite sides, and determine whether each quadrilateral is a trapezoid or parallelogram.

   a) (3, 10), (9, 13), (12, 10), (6, 7)

   b) (10, 5), (5, 6), (3, 2), (8, 1)

   c) (13, 3), (16, 13), (19, 11), (16, 1)

   d) (2, 3), (4, 10), (7, 8), (6, 1)

   e) (6, 8), (15, 8), (14, 14), (4, 14)

   f) (2, 2), (10, 2), (9, 6), (4, 6)

2. How are the opposite angles of a parallelogram related? Compare with Problem 2 of Exercises 4–6.

3. How are two consecutive angles of a parallelogram related? Compare with Problem 2 of Exercises 4–6.

4. In parallelogram $ABCD$, diagonal $\overline{AC}$ can be viewed as a transversal of each pair of opposite sides.

   a) Why is $\angle CAD \cong \angle ACB$?

   b) Why is $\angle ACD \cong \angle CAB$?

   c) Why is $\triangle ADC \cong \triangle CBA$?

   d) How are the opposites $\overline{AB}$ and $\overline{CD}$ related? $\overline{AD}$ and $\overline{BC}$?

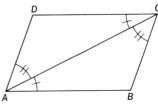

5. In quadrilateral $ABCD$, $\overline{AB} \cong \overline{CD}$, and $\overline{BC} \cong \overline{AD}$.

   a) Why is $\triangle ABC \cong \triangle CDA$?

   b) Why is $\angle BCA \cong \angle DAC$?

   c) Why is $\overleftrightarrow{BC} \parallel \overleftrightarrow{AD}$?

   d) Why is $ABCD$ a parallelogram?

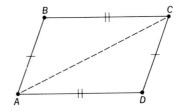

6. In quadrilateral $EFGH$, $\overline{EH} \cong \overline{FG}$, and $\overleftrightarrow{EH} \parallel \overleftrightarrow{FG}$.

   a) Why is $\triangle FGE \cong \triangle HEG$?

   b) Why is $EFGH$ a parallelogram?

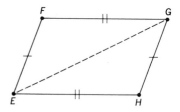

7. Exercises 4, 5, and 6 suggest two other possible definitions of a parallelogram. Formulate these definitions.

8. In parallelogram $ABCD$, P is the intersection of the diagonals.

   a) Why is $\triangle CBP \cong \triangle ADP$?

   b) Why do the diagonals bisect each other?

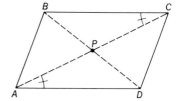

9. In quadrilateral $EFGH$, the diagonals bisect each other.

   a) Why is $\triangle FPE \cong \triangle HPG$?

   b) Why is $EFGH$ a parallelogram?

10. Exercises 8 and 9 suggest another possible definition of a parallelogram. (Explain *why* these exercises suggest a new definition.) Formulate this definition.

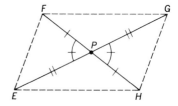

11. Think of two lines in space by visualizing the paths of two jet planes.

    a) If these paths are everywhere the same distance apart what can you say about them?

    b) If the paths are skew will there be two points, one on each line, that are closer to each other than any other such pair of points?

    c) Define *distance between skew lines*.

    d) Try to argue that the shortest segment connecting two skew lines is perpendicular to both lines.

12. Since a rhombus is a parallelogram, its diagonals bisect each other. What other special property have the diagonals of a rhombus? Give a proof.

13. If a parallelogram has one right angle, what can you assert?

14. Of course the diagonals of a rectangle bisect each other. (Why?) What else is true for the diagonals of a rectangle?

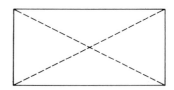

15. A square is a parallelogram, a rhombus, and a rectangle. What can you say say about the diagonals of a square?

16. Two squares with sides of 6 in. overlap as shown, with a vertex of one at the center of the other. What is the area of the shaded part?

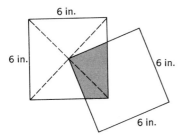

4–9 Parallelepipeds

The *surface* of the solid shown below is a *prism*. To construct a prism, consider two parallel planes π and π', a polygon in π, and a line l intersecting both planes. Consider the lines through vertices of the polygon in π, parallel to l. These lines intersect π' in points which determine a polygon congruent to the polygon in π. In the figure, the polygons $ABCD$ and $A'B'C'D'$ and their interiors are the *bases* of the prism. Each parallelogram like $ABB'A'$ and its interior is a *face* of the prism. (Why is $ABB'A'$ a parallelogram?) The segments $\overline{AA'}$, $\overline{BB'}$, ... are *lateral edges* of the prism. Any segment such as $\overline{EE'}$ with endpoints on the sides

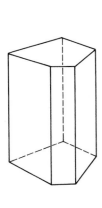

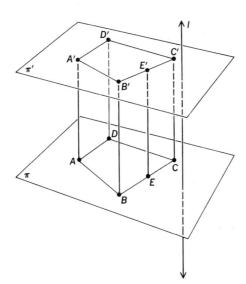

of the bases and parallel to l is called an *element* of the parallelepiped. If the line l is perpendicular to π, the prism is a *right* prism. If the polygon in π is a parallelogram, the prism is called a *parallelepiped*.

Exercises 4–8

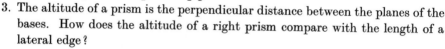

1. A right triangular prism is shown.
   a) Name the bases.
   b) Name the lateral edges.
   c) Name the faces.
   d) What type of figure is each face?
2. Is a cube a prism? A parallelepiped?
3. The altitude of a prism is the perpendicular distance between the planes of the bases. How does the altitude of a right prism compare with the length of a lateral edge?
4. A *horizontal cross section* of a prism is the intersection of the prism with a plane parallel to the bases. How is a cross section related to the bases?

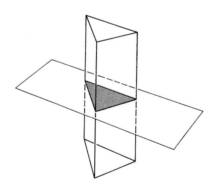

4–10 An Important Theorem

This section deals with a theorem which has far reaching consequences.

> **The segment connecting midpoints of two sides of a triangle is parallel to and half as long as the third side.**

We sketch the proof of this theorem below. In $\triangle ABC$, M and N are midpoints of sides $\overline{AC}$ and $\overline{BC}$ respectively. We show that $\overleftrightarrow{MN} \parallel \overleftrightarrow{AB}$ and $\overline{MN}$ is half as long as $\overline{AB}$. Draw l through B parallel to $\overleftrightarrow{AC}$ and extend $\overline{MN}$ to intersect l at Q.

1. $\triangle CNM \cong \triangle BNQ$ by ASA. (Explain.)
2. $\overline{NM} \cong \overline{NQ}$ and $\overline{MC} \cong \overline{QB}$. (Why?)
3. $\overline{AM} \cong \overline{BQ}$, so $ABQM$ is a parallelogram. (Why?)
4. $\overleftrightarrow{MN} \parallel \overleftrightarrow{AB}$ and $\overline{MN}$ is half as long as AB. (Why?)

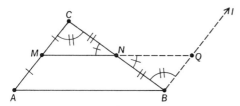

From the theorem above it follows at once that a line through the midpoint of one side of a triangle and parallel to a second side bisects the third side. How would you make this argument? We know that *if* N is the midpoint of $\overline{CB}$, then $\overleftrightarrow{MN} \parallel \overleftrightarrow{AB}$. We also know that there is *only one* line through M parallel to $\overleftrightarrow{AB}$. Complete the argument.

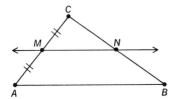

Exercises 4–9

1. Explain why both $BLMN$ and $LCMN$ are parallelograms.
2. Show that $\triangle ANM \cong \triangle NBL \cong \triangle MLC \cong \triangle LMN$.
3. Why is the area of $\triangle ANM$ one-fourth of the area of $\triangle ABC$?

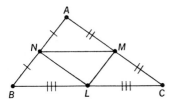

4. Let $ABCD$ be any quadrilateral. Form a second quadrilateral $PQRS$ by connecting the midpoints of the sides as shown. Show that $PQRS$ is a parallelogram. [*Hint:* Compare $\overline{RS}$ with the diagonal $\overline{DB}$.]

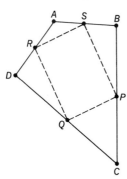

5. Does the theorem suggested by Exercise 4 hold for a nonconvex quadrilateral? For a quadrilateral in space whose four vertices are not coplanar?

6. A triangular pyramid V-ABC is shown. Points M, N, P, Q are midpoints of the indicated sides.

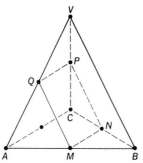

   a) What can you prove about quadrilateral $MNPQ$?
   b) How is (a) related to Exercise 5?
   c) Is line $\overleftrightarrow{VB}$ parallel to the plane of $MNPQ$?
   d) What type of figure is $ABCV$? Answer Exercise 5 again.

7. Points M, N, Q, R, S, and T are midpoints of the edges of pyramid V-ABC. What is the relation between $\triangle MNQ$ and $\triangle RST$?

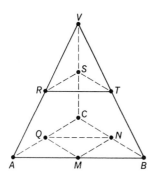

8. Pyramid $V\text{-}ABCD$ has a square base, and M, N, P, Q, R, S are midpoints of the indicated edges.

   a) What type of polygon is $MNPQ$?

   b) What type of polygon is $PQRS$?

   c) How are planes PQR and ABC related?

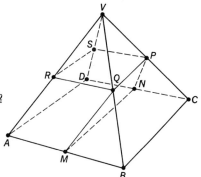

9. In the figure $V\text{-}ABCD\text{-}W$ is an octahedron, and M, N, P, Q, R, S, T, U are midpoints of the indicated segments. Explain why each lateral face of the inscribed solid is a parallelogram.

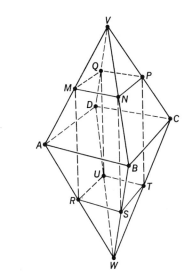

10. Let $\overline{BB'}$ and $\overline{CC'}$ be medians of $\triangle ABC$ intersecting at G. Let L and M be midpoints of $\overline{BG}$ and $\overline{CG}$ respectively.

   a) Why is $\overleftrightarrow{LM} \parallel \overline{B'C'}$?

   b) Why is $\overleftrightarrow{C'L} \parallel \overleftrightarrow{B'M}$?

   c) Why is $C'LMB'$ a parallelogram?

   d) Why is $\overline{B'G} \cong \overline{GL}$ and $\overline{C'G} \cong \overline{GM}$?

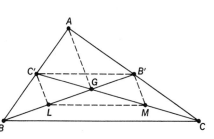

   e) As a result of (a), (b), (c), and (d), we can say that any two medians of a triangle intersect in a point two-thirds of the way from the vertex to the midpoint of the opposite side. How does this result show that the three medians of a triangle are concurrent?

4–11 Volume and Surface Area of Pyramids and Prisms

In Chapter 2, we considered the volume and surface area of solids. In some cases, where the figures were "nice," we could compute the volume and surface area exactly by counting cubes; in other situations we estimated the volume and surface areas by comparing the solid with a cube or rectangular right prism. In this section, we will develop formulas for finding the volume and surface area of prisms and pyramids. When we write the volume of a polyhedron, we really refer to the volume of the region of space bounded by the polyhedron.

Mathematicians speak of two types of surface areas, total and lateral. The *total surface area* is the area of lateral faces plus the area of the base(s), whereas *lateral area* refers to the area of the lateral faces only. We shall continue to use surface area for total surface area.

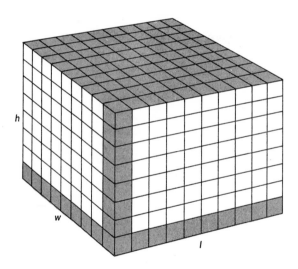

Exercises 4–10

1. In the prism,
    a) How many cubes are in the bottom layer?
    b) Count the number of cubes in the solid.
    c) Explain why the volume of the rectangular solid is $w \cdot l \cdot h$.
    d) What is the surface area of the solid?

2. The figure at the top of the following page shows a solid with height h, length l, and width w. Using the dotted lines, explain why the volume of this region is found in the same way as the volume for the solid in Exercise 1, $V = w \cdot l \cdot h$.

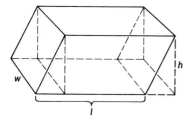

3. The volume of a box (right prism) is $w \cdot l \cdot h$.

   a) By what multiple is the volume increased if each dimension is doubled?

   b) By what multiple is the surface area increased if each dimension is doubled?

   c) By what multiple is the volume increased if each dimension is tripled?

4. Can you compute the surface area of the figure in Exercise 2 if l, w, and h are known?

5. Imagine a right prism (a) made of large slabs of steel. Slide these slabs to form figure (b) or figure (c).

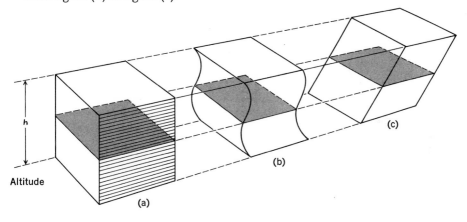

   a) Is the altitude of each figure the same?

   b) Are the cross sections of each figure the same?

   c) Is the volume of each figure the same?

   d) Explain why the volume of any prism is given by the formula $V = B \cdot h$, where B is the area of the base.

The principle involved in the preceding problem is called *Cavalieri's Principle*. We state it below.

If two solids have equal altitudes and if cross sections of these solids formed by planes parallel to the bases and at the same distance from the bases have equal areas, then the solids have equal volumes.

6. Construct a prism and a pyramid with the same base and the same altitude. (Use stiff cardboard.) Remove the top of the prism and the base of the pyramid. Use sand to compare the volume of the pyramid and the prism. What is the formula for the volume of a pyramid?

7. Find the volume of prisms (a) through (h).

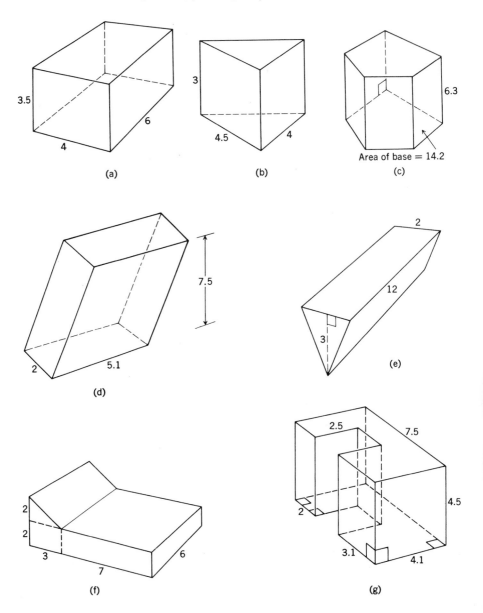

(a)   (b)   (c)

Area of base = 14.2

(d)   (e)

(f)   (g)

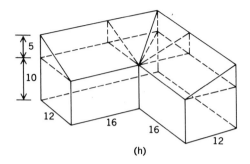

(h)

8. Find the surface area of the solids a, b, e, f, g, and h of Exercise 7. What is the surface area of the solid in Exercise 7(c) if the perimeter of a base is 16? Write a formula for the lateral surface area of a prism.

9. Find the volume of a regular square pyramid whose faces are equilateral triangles and whose base has an edge of 12. Also find the lateral surface area (surface area not including the base).

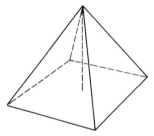

10. If two regular square pyramids whose lateral faces are equilateral triangles are placed base to base, the resulting solid is called a regular octahedron. Show that the volume V of the regular octahedron whose edge is e, is given by $V = \frac{1}{3}\sqrt{2e^3}$. What is the surface area?

$V = \frac{1}{3}\sqrt{2}\ e^3$

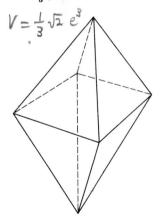

11. Describe a technique for finding the volume of a rock which is irregularly shaped.

12. Think of a box with dimensions l, w, h. Now think of each dimension increased by 2 so that the new dimensions are $(l + 2)$, $(w + 2)$, and $(h + 2)$.

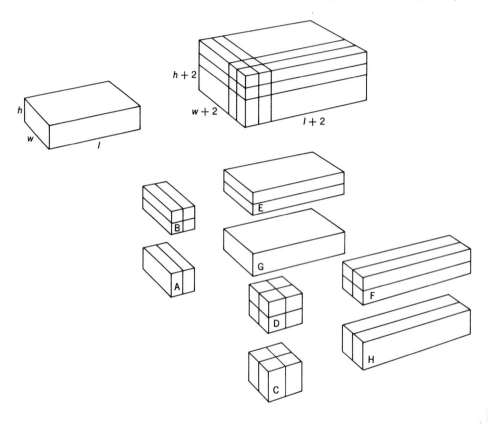

The figure shows this "new box" broken up into eight solids. The volume of the new box can now be worked out. Match each part of the equation with one of the eight regions.

$$V = (l + 2) \cdot (w + 2) \cdot (h + 2)$$
$$= l \cdot (w + 2) \cdot (h + 2) + 2(w + 2) \cdot (h + 2)$$
$$= (l \cdot w + 2 \cdot l) \cdot (h + 2) + (2w + 4) \cdot (h + 2)$$
$$= (l \cdot w \cdot h) + (2 \cdot l \cdot w) + (2 \cdot l \cdot h) + (4 \cdot l) + (2 \cdot w \cdot h) + (4 \cdot w) + (4 \cdot h) + 8$$

$$\begin{array}{cccccccc} \uparrow & \uparrow & \uparrow & \uparrow & \uparrow & \uparrow & \uparrow & \uparrow \\ 1 & 2 & 3 & 4 & 5 & 6 & 7 & 8 \end{array}$$

13. a) Which of these figures can be traced without lifting your pencil from the paper or retracing?

b) A vertex is *odd* if there are an odd number of segments meeting at the vertex. How many odd vertices are there for the curves that can be traced in this manner? (Intersections of diagonals are not considered vertices.)

c) Draw other figures which can be traced in this manner.

d) Draw figures which cannot be traced in this manner.

e) Make a conjecture about the number of odd and even vertices for a "traceable" figure.

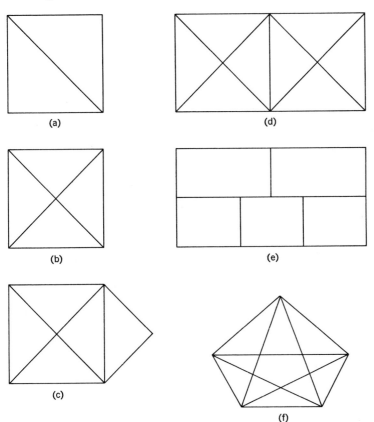

4–12 Sum of the Angles of a Triangle

We defined a *sum* of two angles A and B as a third angle formed from two adjacent angles, one congruent to $\angle A$, the other to $\angle B$. Note that if we picture the

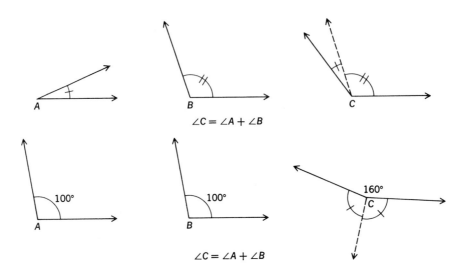

$$\angle C = \angle A + \angle B$$

$$\angle C = \angle A + \angle B$$

sum of two 100° angles as above we get the strange result that their sum is a 160° angle! This situation is slightly embarrassing but it causes no difficulty. In advanced mathematics it is customary to define *directed angles* and to speak of angles of 200°, 1000°, −450°, etc. This procedure is too complicated for our purposes. We shall avoid difficulty by considering the sums of the *measures* of angles. That is, we shall remark that the sum of the *measures* of the angles of a square is 360° Of course, in the geometric sense of "adding" the angles of a square one might claim that the sum of these four angles is a *zero* angle. (Explain.) When we speak of "adding angles" in the sequel we shall mean *adding the real numbers which are the measures of the angles.*

One of the most dramatic theorems of Euclidean geometry is that for every triangle, the sum of its angles is a straight angle. This is a theorem peculiar to Euclidean geometry. For its proof Euclid's parallel postulate is needed. In the exercises that follow we present two different proofs and explore some of the consequences of the theorem. One proof lends itself nicely to a paper-folding experiment.

Exercises 4–11

1. Points M, N are midpoints of sides $\overline{AC}$ and $\overline{BC}$ of $\triangle ABC$. "Fold" the three corners of the triangle on the dotted lines shown. Then *all three vertices, A, B, C, fall upon P, and the three angles, $\angle A$, $\angle B$, $\angle C$, fit together to form a straight angle.*

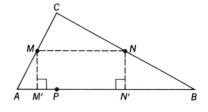

2. Exercise 1 is the paper-folding experiment. Cut out a paper triangle and make this experiment. The italicized phrases in Exercises 1 indicate what must be proved in order to provide logical support for the conjecture suggested by the experiment. We indicate below how the proof can be completed.

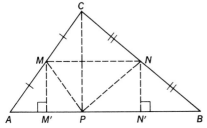

a) If M' and N' are the feet of perpendiculars to AB from M and N, then $\overline{M'P} \cong \overline{AM'}$, and $\overline{BN'} \cong \overline{N'P}$. (Why?)

b) Part (a) shows that when we "fold" on $\overline{MM'}$ and $\overline{NN'}$, then A and B both fall upon P. (Why?)

c) In the figure, $\triangle AMM' \cong \triangle PMM'$ and $\triangle BNN' \cong \triangle PNN'$. (Why?)

d) $\triangle MCN \cong \triangle MPN$. (Why?)

e) $\overset{\leftrightarrow}{CP} \perp \overset{\leftrightarrow}{MN}$. (Why?)

f) Parts (d) and (e) show that when we "fold" on $\overline{MN}$, $\angle C$ falls upon $\angle MPN$. Consequently, using (a), (b), (c), (d), and (e) together, we see that $\angle A$, $\angle B$, and $\angle C$ form a straight angle. (Explain.)

g) The $\triangle ABC$ above has all acute angles. Explain how you would perform the paper-folding experiment if the triangle is a right triangle or an obtuse triangle.

3. A cruder experiment that does not suggest a supporting geometric proof simply consists of ripping off the corners of a triangle and placing the angles together. Perform this experiment.

4. Perhaps the simplest and most satisfying proof of our angle-sum theorem for triangles is based upon properties of parallels cut by a transversal. In the accompanying figure $l \parallel \overset{\leftrightarrow}{AB}$.

a) Why are the indicated angles congruent?

b) How does this establish that the angle sum is 180°?

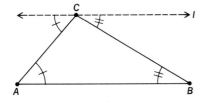

5. It is a simple step to theorems on the angle-sum for convex polygons other than triangles.

   a) What is the sum of the degree measures of the angles labeled 1 through 6? What is the sum of the angles of any convex quadrilateral?

   b) What is the sum of the angles of any convex pentagon?

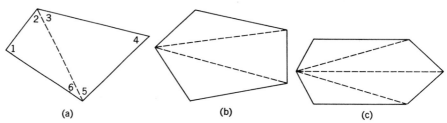

(a)   (b)   (c)

   c) What is the sum of the angles of any convex hexagon?

   d) What is the sum of the angles of any convex n-gon (polygon of n sides)?

6. The figures suggest an alternative to the proof of Exercise 5 that the sum of the angles of any convex n-gon is $(n - 2)$ straight angles.

   a) What is the sum of angles 1 through 12? Of angles 9, 10, 11, and 12? Of angles 1 through 8? Of the angles of the quadrilateral?

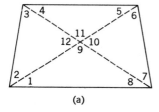

(a)

   b) As in (a), compute the sum of the angles of the pentagon.

   c) The figure suggests a convex n-gon. As in (a) and (b) compute the sum of its n angles.

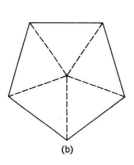

(b)

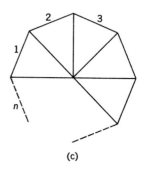

(c)

7. A seven-sided convex polygon is pictured. Its angles are labeled 1 through 7, and its exterior angles are correspondingly labeled 1′ through 7′.

   a) Give, in terms of the straight angle as the unit, the sum of all fourteen angles, the seven interior and the seven exterior angles.

   b) From earlier exercises, what is the sum of the seven interior angles?

   c) What is the sum of the seven exterior angles?

   d) What is the sum of the n exterior angles for any convex n-gon?

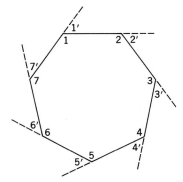

8. All the angles of an *equiangular* polygon are congruent. Since we know the sum of the angles for any convex polygon, we can compute the measure of *each* angle for any convex equiangular polygon. Explain this assertion, and compute the measure of each angle in a convex equiangular polygon of (a) 4 sides, (b) 5 sides, (c) 6 sides, (d) 10 sides, (e) 100 sides, (f) n sides.

9. Triangle ABC has a right angle at C. Measures of certain angles are indicated.

   a) Find the degree measure for each of $\angle ECB$, $\angle ABC$, and $\angle CDE$.

   b) If C is not a right angle, can you still compute the sum of the measures of the three angles of (a)?

10. What is the measure of each vertex (1, 2, 3, 4, 5) angle of a regular five-pointed star? [*Hint: ABCDE* is a *regular* pentagon with all sides congruent and all angles congruent.]

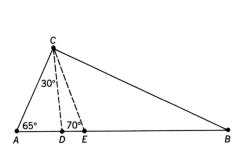

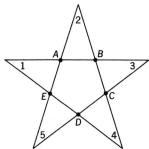

11. What is the sum of the measures of the five vertex angles of the "irregular" five-pointed star?

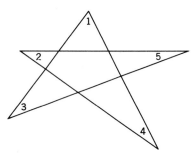

12. In the accompanying figures, the indicated congruence relations hold.
    a) What is true for $\angle C$ and $\angle C'$?
    b) Why is $\triangle ABC \cong \triangle A'B'C'$?
    c) State a congruence theorem for triangles suggested by this exercise.

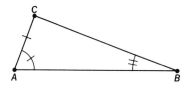 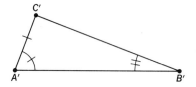

13. Relate Exercise 12 to the theorem:

    Two right triangles are congruent if the hypotenuse and an acute angle of one are congruent respectively to the hypotenuse and an acute angle of the other.

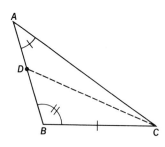 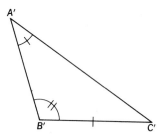

In Exercise 12 above, the congruence theorem is angle, angle, side (AAS). We present a proof of the theorem. In $\triangle ABC$ and $\triangle A'B'C'$,
$$\angle A \cong \angle A', \ \angle B \cong \angle B', \text{ and } \overline{BC} \cong \overline{B'C'}.$$

To prove that $\triangle ABC \cong \triangle A'B'C'$:

If $\overline{AB} \cong \overline{A'B'}$, then $\triangle ABC \cong \triangle A'B'C'$ by ASA. Either $\overline{AB} \cong \overline{A'B'}$ or $\overline{AB} \ncong \overline{A'B'}$. Assume $\overline{AB} \ncong \overline{A'B'}$; then either $\overline{AB} > \overline{A'B'}$ or $\overline{AB} < \overline{A'B'}$. If $\overline{AB} > \overline{A'B'}$, take D on $\overline{AB}$ such that $\overline{A'B'} \cong \overline{BD}$. Then $\triangle DBC \cong \triangle A'B'C'$. (Why?) Hence $\angle BDC \cong \angle B'A'C'$, and therefore $\angle BDC \cong \angle BAC$. (Why?) This a contradiction. (Explain.) Since the assumption that $\overline{AB} \ncong \overline{A'B'}$ leads to a contradiction, it must be that the assumption is false. The only other possibility is that $\overline{AB} \cong \overline{A'B'}$ and this implies $\triangle ABC \cong \triangle A'B'C'$. (A similar argument holds if $\overline{AB} < \overline{A'B'}$.) Note that the proof is the same if $\overline{AC} \cong \overline{A'C'}$ rather than $\overline{BC} \cong \overline{B'C'}$.

The most systematic attempt to base a proof of the parallel postulate upon Euclid's other axioms was made by the Italian mathematician Saccheri in the 17th century. We describe some of his efforts below. The figure shown is called a *Saccheri quadrilateral*. The segment $\overline{AB}$ is called the *base* and $\overline{CD}$ the *summit*.

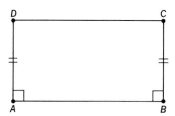

The other sides $\overline{AD}$ and $\overline{BC}$ are congruent to each other and perpendicular to the base. Angles D and C are called *summit angles*. Saccheri used this quadrilateral in his investigations of the parallel postulate. It is easy to show (without using the parallel postulate) that

1. The summit angles are congruent ($\angle D \cong \angle C$).
2. The sum of the angles of each triangle is the sum of the summit angles of *some* Saccheri quadrilateral.

We sketch proofs of these two assertions below, leaving details to the reader.

Proof of 1. In the Saccheri quadrilateral $ABCD$, let M and N be midpoints for the base and summit as shown.

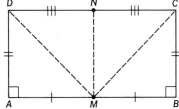

a) $\triangle AMD \cong \triangle BMC$. (Why?)
b) $\triangle DMN \cong \triangle CMN$. (Why?)
c) $\angle ADN \cong \angle BCN$. (Why?)

Proof of 2. With M, N the midpoints of the sides $\overline{AC}$ and $\overline{BC}$ as shown and $\overline{AD}$, $\overline{CF}$, and $\overline{BE}$ perpendicular to $\overleftrightarrow{DE}$, we have the following relationships:

a) $\triangle AMD \cong \triangle CMF$ and $\triangle BEN \cong \triangle CFN$. (Why?)

b) $DEBA$ is a Saccheri quadrilateral. (Why?)

c) $\angle ABC + \angle BCA + \angle CAB$ is the sum of the summit angles, $\angle DAB$ and $\angle EBA$. (Why?)

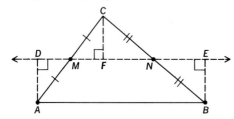

From Proof 1 and Proof 2 above, it follows that if one of the summit angles of every Saccheri quadrilateral is a right angle, then the sum of the angles of every triangle is two right angles, as stated in Euclidean geometry. If the summit angles of every Saccheri quadrilateral are acute, then every triangle has an angle sum less than 180°. If the summit angles of every Saccheri quadrilateral are obtuse, then every triangle has an angle sum of more than 180°. Saccheri was able to prove that no Saccheri quadrilateral has an obtuse summit angle. He showed also that if there is *one* Saccheri quadrilateral with a right summit angle, then *every* Saccheri quadrilateral has a right summit angle. The assumption that there is one Saccheri quadrilateral with a right summit angle is equivalent to the Parallel Postulate of Euclid. Saccheri tried desperately to prove that no Saccheri quadrilateral has an acute summit angle. He failed, and died never knowing the implications of his failure. Years after the labors of Saccheri had been drawn to a conclusion, Lobachevski published his work on hyperbolic non-Euclidean geometry. In this geometry every Saccheri quadrilateral has acute summit angles; every triangle has an angle sum less than 180°; very small triangles have angle sums very near 180°; and the larger the triangle, the smaller its angle sum.

Problem. Use the figure and prove that *if* all triangles have the same angle sum, then that sum is 180°. [*Hint:* Let D be any point between B and C, and consider the sum of the angles for each of the triangles, ABC, ABD, and ADC.]

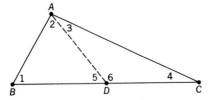

5 Algebraic Structures I

5–1 Introduction

We now turn our attention to algebra. In this chapter, Part I, we stress algebraic ideas that have clearcut geometric implications. In Chapter 6, Part II, the algebra will have little connection with geometry.

In these two chapters we have three main objectives. In your contact with school algebra, you probably thought of algebra as generalized arithmetic. In this algebraic work you calculated with letters which represented numbers. This interpretation of algebra as a study of the properties of the familiar number systems is most important for school mathematics, and so our first objective is to describe the structure of these number systems, the whole numbers, integers, rationals, and real numbers.

A second objective is to show how algebraic techniques can aid in the study of geometry. Algebra and geometry are brought together by the invention of coordinate systems. A coordinate system enables us to use numbers to name points, and then our skills in calculating with numbers enable us to discover the geometric properties of sets of points.

The third objective is to point out by examples that algebra is far more than a study of number systems. An important recent development in algebra is that mathematicians have invented many algebraic systems whose elements are not numbers. We shall introduce some of these systems in this chapter and note their geometrical significance.

5–2 The Structure of Our Number System

One task of algebra is to give a clear description of the properties possessed by various mathematical systems. Among these are our familiar number systems all of which are subsystems of the system of complex numbers.

1. The whole numbers: $W = \{0, 1, 2, \ldots\}$.

2. The nonnegative rationals: $Q^+ = \{0, 1, 2, \ldots; \frac{1}{2}, \frac{1}{3}, \ldots\}$.

3. The integers: $I = \{\ldots -2, -1, 0, 1, 2, \ldots\}$.

4. The rationals: $Q = \{0, \pm 1, \pm 2, \ldots; \pm \frac{1}{2}, \pm \frac{1}{3}, \ldots\}$.

5. The real numbers: R consists of all the rationals Q, and certain irrational numbers.

6. The complex numbers: C includes R as well as certain numbers labeled "imaginary."

In this section we consider the basic properties that these number systems possess, emphasizing both their similarities and the differences that distinguish them from each other. The first three of these, W, Q^+, and I, are the important ones in the elementary school.

We think of each of these number systems as a set of elements with two operations, addition and multiplication. Addition and multiplication are called *binary* operations on these sets, since each operation associates with each ordered *pair* of elements of one of these sets a single element of that set. For example,

$$2 \in W, \quad 3 \in W, \quad \text{and} \quad 2 + 3 \in W, \quad 2 \cdot 3 \in W,$$

$$\sqrt{2} \in R, \quad \sqrt{5} \in R, \quad \text{and} \quad \sqrt{2} + \sqrt{5} \in R, \quad \sqrt{2} \cdot \sqrt{5} \in R.$$

In all these sets the addition and multiplication operations possess the commutative and associative properties. For example, in W,

Comm $+$: $\forall a, b \in W, a + b = b + a$.

Assoc $+$: $\forall a, b, c \in W, (a + b) + c = a + (b + c)$.

Comm $\cdot$: $\forall a, b \in W, a \cdot b = b \cdot a$.

Assoc $\cdot$: $\forall a, b, c \in W, (a \cdot b) \cdot c = a \cdot (b \cdot c)$.

In each of these number systems each of the operations $+$ and $\cdot$ has a *neutral* or *identity* element. For addition this neutral element is 0; for multiplication, 1. We refer to the additive property of 0 (Add 0) and the multiplicative property of 1 (Mult 1). For example, in Q,

Add 0: $\forall r \in Q, r + 0 = 0 + r = r$.

Mult 1: $\forall r \in Q, r \cdot 1 = 1 \cdot r = r$.

In each of these number systems the multiplication operation is *distributive* over addition. For example, in R,

Dist: $\forall x, y, z \in R, x \cdot (y + z) = x \cdot y + x \cdot z$.

Also, in all these systems addition and multiplication obey *cancellation* laws. In W,

Cancel $+$: $\forall a, b, c \in W, a + b = a + c \Rightarrow b = c.$

Cancel $\cdot$: $\forall a, b, c \in W, a \cdot b = a \cdot c$ and $a \neq 0 \Rightarrow b = c.$

The nine properties described above are enjoyed by each of the six number systems $W, Q^+, I, Q, R,$ and C.

Comm $+$	Comm $\cdot$
Assoc $+$	Assoc $\cdot$
Cancel $+$	Cancel $\cdot$
Add 0	Mult 1
Dist	

Common Properties of $W, Q^+, I, Q, R,$ and C.

Now we focus attention upon differences in these systems. Note that each number in I has an *additive inverse* or *negative*,

$$\forall a \in I, \exists -a \in I$$

such that

$$a + (-a) = 0.$$

Each nonzero number in Q has a *multiplicative inverse* or *reciprocal*,

$$\forall r \in Q, r \neq 0, \exists \frac{1}{r} \in Q$$

such that

$$r \cdot \frac{1}{r} = 1.$$

The number systems R and C also have these properties, however,

(i) I has additive inverses but not multiplicative inverses.

(ii) Q^+ has multiplicative inverses but not additive inverses.

(iii) W has neither additive nor multiplicative inverses.

Some new terminology is convenient. Because in I each element has an inverse relative to the operation $+$, we say that I is a group relative to addition. Similarly, Q^+ (excluding the number 0) is a group relative to multiplication. I is not a group relative to multiplication, since it lacks inverses. We do call I a *semigroup* (half group) relative to multiplication. W is a semigroup relative to both addition and multiplication. The systems $Q, R,$ and C are all called *fields*.

The concepts of group and field are so important that we formulate definitions below.

<table>
<tr>
<td>

A set is a group relative to a binary operation if
   a) The operation is associative,
   b) The operation has an identity element, and
   c) Each element has an inverse relative to the operation.

</td>
<td>

A set with two binary operations, addition and multiplication, is a field if
   a) The set is a commutative group relative to addition,
   b) The nonzero elements form a commutative group relative to multiplication, and
   c) Multiplication is distributive over addition.

</td>
</tr>
<tr>
<td style="text-align:center">Definition of Group</td>
<td style="text-align:center">Definition of Field</td>
</tr>
</table>

The terminology (group, field) that we have introduced facilitates the description of differences between W, I, and Q^+. However, Q, R, and C are all fields, and so we need to develop other concepts in order to pinpoint differences between these number systems. One of the simplest ways to identify these differences is to consider the task of solving equations. In each of these three fields we can solve every *linear* equation, that is every equation of the form $ax + b = c$; a, b, c in the field, $a \neq 0$. We simply add the negative of b to both sides of the equation and multiply by the reciprocal of a:

$$ax + b = c \Rightarrow x = \frac{1}{a}\cdot[c + (-b)].$$

But as soon as we consider equations of second degree, differences in these systems become apparent.

i) The equation $x^2 = 2$ has no root in Q, but does have two roots ($\sqrt{2}$ and $-\sqrt{2}$) in R and in C.

ii) The equation $x^2 = -1$ has no root in R (and hence none in Q, since $Q \subset R$), but has two roots in C ($\sqrt{-1}$ and $-\sqrt{-1}$, usually abbreviated as i and $-i$).

The fields Q and R are *ordered* fields relative to an order relation *less than* denoted by the symbol $<$. This relation $<$ has the following properties,

i) *Trichotomy:* $\forall a, b \in R$, exactly one of $a < b$, $a = b$, $b < a$ holds.

ii) *Transitivity:* $\forall a, b, c \in R$, $a < b$ and $b < c \Rightarrow a < c$.

iii) *Monotony:* $\forall a, b, c \in R$.

    1) $a < b \Rightarrow a + c < b + c$,

    2) $a < b$ and $0 < c \Rightarrow ac < bc$.

The complex field of numbers is not an ordered field; that is, there is no such relation $<$ in C with the properties of Trichotomy, Transitivity, and Monotony. Of course W and I are ordered in the sense that the three properties above hold for them also.

Some properties of Q and R can be described in terms of the order relation. For example, Q is *dense in itself*. This means that between any two rational numbers there lies a third,

$$\forall a, b \in Q, a < b, \exists c \in Q \qquad \text{such that} \qquad a < c < b.$$

The set R of real numbers is dense in itself also. More interestingly, Q is dense in R. That is,

$$\forall x, y \in R, x < y, \exists r \in Q \qquad \text{such that} \qquad x < r < y.$$

Paradoxically, even though Q is dense in R, there are fewer rational numbers than irrational numbers.

We say that R is a *complete* ordered field. A precise definition of this property of completeness is difficult to formulate, and we shall not do so here. Intuitively you can think of the completeness of the real number system as reflecting the fact that a one-to-one correspondence can be established between R and the points of any line. There are no *holes* in the real number line.

Exercises 5–1

 No

1. In the set I of integers we denote W by I^+ and the negatives of the elements of I^+ by I^-.

$$I^+ = \{0, 1, 2, \ldots\}, \qquad I^- = \{0, -1, -2, \ldots\}.$$

    a) Show that $+$ is an associative binary operation on I^-. (We say that the subset I^- is *closed* under addition.)

    b) Explain why I^- is a semigroup rather than a group relative to addition.

    c) Explain why I^- is not a semigroup relative to multiplication.

2. Let E be the subset of even integers, $E = \{\ldots, -4, -2, 0, 2, 4, \ldots\}$.

    a) Show that E is a group relative to addition.

    b) Is E closed relative to multiplication?

    c) Show that E is not a group relative to multiplication.

3. Let O be the subset of odd integers, $O = \{\ldots, -3, -1, 1, 3, \ldots\}$.

   a) Show that O is not a group relative to addition.

   b) Show that O is not a group relative to multiplication.

4. Let Q^+ be the set of positive rational numbers. Show that Q^+ is a group relative to multiplication.

5. In the set I define a binary operation $\oplus$ by agreeing that

   $a \oplus b = 2a + b.$

   a) Show that $\oplus$ is not commutative.

   b) Show that $\oplus$ is not associative.

   c) Show that $\oplus$ does not have an identity element.

6. In the set I define a binary operation $\oplus$ by agreeing that

   $a \oplus b = 2a + 2b.$

   a) Is $\oplus$ commutative?

   b) Is $\oplus$ associative?

   c) Is ordinary multiplication distributive over $\oplus$?

   d) Has $\oplus$ an identity element?

7. In the set I define two binary operations $\oplus$ and $\odot$ by agreeing that $a \oplus b = a + b - 1$; $a \odot b = ab - a - b + 2$.

   Compute the following:

   a) $3 \oplus 4$ b) $4 \oplus 3$ c) $3 \odot 4$ d) $4 \odot 3$

   e) $(1 \oplus 2) \oplus 3$ f) $1 \oplus (2 \oplus 3)$ g) $3 \oplus (-1)$

   h) $5 \oplus (-3)$ i) $(-3) \oplus 1$ j) $(2 \odot 3) \odot 4$

   k) $2 \odot (3 \odot 4)$ l) $2 \odot (3 \oplus 4)$ m) $(2 \odot 3) \oplus (2 \odot 4)$

8. With $\oplus$ and $\odot$ the binary operations of Exercise 7,

   a) Is $\oplus$ commutative?

   b) Is $\oplus$ associative?

   c) Has $\oplus$ an identity element?

   d) What is the inverse of 5 relative to $\oplus$?

   e) Is $\odot$ commutative?

   f) Is $\odot$ associative?

   g) Is $\odot$ distributive over $\oplus$?

   h) Has $\odot$ an identity element?

9. Consider the set of all rational numbers whose lowest terms representations have even denominators.

   a) Is this set closed under addition?

   b) Is this set closed under multiplication?

   c) Is this set a group relative to either addition or multiplication?

10. Consider the set of all rational numbers whose lowest terms representations have odd denominators.

    a) Is this set closed under addition?

    b) Is this set closed under multiplication?

    c) Is this set a group relative to either addition or multiplication?

11. In the field of real numbers let a and b be any two real numbers and $\bar{a}$, $\bar{b}$ their additive inverses. Show that $\bar{a} + \bar{b}$ is the additive inverse of $a + b$.

12. In the field of complex numbers let r and s be any two nonzero numbers, and let r^{-1} and s^{-1} be their multiplicative inverses. Show that $r^{-1} \cdot s^{-1}$ is the multiplicative inverse of $r \cdot s$.

13. Consider the set of all rational numbers whose lowest terms representations have both odd denominators and odd numerators.

    a) Is this set closed under addition?

    b) Is this set closed under multiplication?

    c) Is this set a group relative to either addition or multiplication?

14. Write an expression representing the additive inverse of each number below in the field C of complex numbers:

    a) -3
    b) $\dfrac{5}{-4}$
    c) $\dfrac{-3}{-7}$
    d) $2 + \frac{3}{5}$

    e) $\sqrt{5}$
    f) $3 - \sqrt{2}$
    g) $\dfrac{1}{\sqrt{7}}$
    h) $\dfrac{1}{\sqrt{2}} + \dfrac{1}{\sqrt{3}}$

    i) $i - 1$
    j) $\dfrac{1}{i}$
    k) $4 - \sqrt{-7}$
    l) $\dfrac{3 - i}{i + 5}$

15. Write an expression representing the reciprocal (multiplicative inverse) of each number of Exercise 14 in the field C of complex numbers.

16. In the set of integers I, if we know that for integers a, b, and c

    $$a + b = a + c,$$

    then by adding $-a$ to each side of this equation and using the associative property of $+$ and the property of 0, we can draw a certain conclusion. State the conclusion.

17. In the set of whole numbers W, the nonzero numbers do not have negatives.

Hence we have no obvious way to *prove* that

if $a + b = a + c$,   then $b = c$.

Yet this is a result that we need if we wish to define subtraction in W. Explain this last remark and suggest a way out of our difficulty.

18. In the set Q of rational numbers, if we know for rationals a, b, c that $a \neq 0$, and $a \cdot b = a \cdot c$, then by multiplying each side of the equation by $1/a$ and using the associative property of $\cdot$ and the property of 1, we can draw a certain conclusion. Show that this is so and state the conclusion.

19. In the set of integers I, the integers other than 1 and -1 have no reciprocals. Hence we have no obvious way to *prove* that

if $a \neq 0$ and $a \cdot b = a \cdot c$,   then $b = c$.

Yet this is a result that we need if we wish to define division in I. Explain this last remark and suggest a way out of our difficulty.

20. Suppose that you know that in a mathematical system, with an associative binary operation $*$ and an identity element e relative to this operation, there are elements a, b, c such that

$a * b = a * c$ and $b \neq c$.

What can you assert about the element a?

We have not discussed subtraction or division. These operations are defined in terms of addition and multiplication. In I the simplest way to introduce subtraction is via the definition,

i) $\forall a, b \in I, a - b = a + (-b)$.

In Q^+ the simplest way to introduce division is

ii) $\forall a, b \in I, b \neq 0, a \div b = a \cdot \dfrac{1}{b}$.

We can make these definitions because each number in I has a negative and each nonzero number in Q^+ has a reciprocal. However, neither of these definitions is applicable in W, nor can we thus define division in I or subtraction in Q^+. (Why?) Note that Definitions (i) and (ii) can be formulated in any field. (Why?)

In W we are faced with the fact that we cannot define subtraction (or division) so that every number can be subtracted from (or divided into) every other number, and so we make the definitions,

iii) $\forall a, b, c \in W$,   *if* $b + c = a$,   then $c = a - b$,

iv) $\forall a, b, c \in W$,   *if* $b \cdot c = a$ and $b \neq 0$,   then $c = a \div b$.

The properties of subtraction and division are vastly different from those of addition and multiplication. Give several reasons why none of our familiar number systems are groups relative to either subtraction or division.

5–3 Linear Functions

In Section 5–2 we analyzed mathematical systems whose elements are numbers. The operations in these systems are operations on numbers. But in algebra we often concern ourselves with systems whose elements are not numbers. In this section we will consider a set of functions (operations) which has great importance in elementary arithmetic. Our operations in this section will be operations on these functions, not operations on numbers. The idea is really very simple. Children in the early grades, who reason as below, exemplify the ideas we shall develop in this section.

i) To *add* 3 and then *add* 7 is the same as to *add* 10.

ii) To *multiply* by 2 and then by 5 is the same as to *multiply* by 10.

iii) To *add* 4 and then *multiply* by 2 is the same as first *multiplying* by 2 and then *adding* 8.

You have often heard it said that addition and subtraction are *inverse* operations. What this assertion really means is that for each of the *functions* (operations) adding 3, adding 7, ..., there is an *inverse function* subtracting 3, subtracting 7, ...

There are several ways to symbolize the concepts that we are introducing here. Perhaps the simplest technique is to use symbols like A_2 and M_3 for the operations of adding 2 and multiplying by 3, respectively. With this notation the equations below should be self-explanatory.

$$A_2 A_4 = A_6, \qquad M_3 M_5 = M_{15}.$$

Sometimes it will be convenient to use the function notation below:

$$A_3(x) = x + 3, \qquad M_2(x) = 2x, \qquad A_4 M_3(x) = A_4(3x) = 3x + 4;$$
$$M_3 A_4(x) = M_3(x + 4) = 3(x + 4) = 3x + 12 = A_{12} M_3(x).$$

Read these equations as, "A three of x equals $x + 3$," "M two of x equals $2x$," etc. Note that the last calculation justifies our writing

$$M_3 A_4 = A_{12} M_3.$$

That is: For every real number x, if we first add 4 to x and then multiply this sum by 3, the result is the same as if we first multiply x by 3 and then add 12. (We apologize for making you read some of these function expressions from right to left. Why do we do it?)

Observe that in your high-school algebra work, when you wrote

$$(x + 2) + 3 = x + 5, \qquad 2(x + 3) = 2x + 6,$$

you were asserting respectively that

$$A_3A_2 = A_5, \qquad M_2A_3 = A_6M_2.$$

We shall refer to the set of all functions that we can build up from the functions A_x, M_y, where x and y are real numbers, as the set of *linear* functions.

We shall say that A_5 is the *composite* of A_2 and A_3, and the binary operation that combines A_2 and A_3 will be called the *composition* of functions. In the next exercise list we develop properties of this set of linear functions. For convenience, we denote by $\mathscr{A}$, the set of all addition functions A_x for real numbers x; by $\mathscr{M}$, the set of multiplication functions M_y for all real numbers $y \neq 0$; and by $\mathscr{L}$, the set of all functions that can be constructed by composing the addition and multiplication functions.

Exercises 5–2

1. With our notation, $A_rA_s(x)$ means $A_r(x + s) = (x + s) + r = x + (s + r) = A_{s+r}(x) = A_{r+s}(x)$. Hence $A_rA_s = A_{r+s}$.

   a) Show as above that for any two multiplication functions M_r, M_s that $M_rM_s = M_{r \cdot s}$.

   b) Show that each of the sets $\mathscr{A}$ and $\mathscr{M}$ is closed under the operation composition of functions.

   c) Show that in each of the sets $\mathscr{A}$ and $\mathscr{M}$, composition is an associative operation.

   d) Show that in each set $\mathscr{A}$ and $\mathscr{M}$ there is a special function which is the neutral element relative to the composition operation.

   e) Note that $A_{-3}A_3(x) = A_3A_{-3}(x) = A_0(x)$. Show that each element in $\mathscr{A}$ has an inverse as does also each element in $\mathscr{M}$. What is the inverse of $A_{4/5}$? A_{-7}? $M_{2/3}$? M_{-5}?

   f) Is composition a commutative operation in each of the sets $\mathscr{A}$ and $\mathscr{M}$?

2. Explain why the results listed in Exercise 1 justify the assertions below.

   i) The set of addition operations $\mathscr{A}$ is a commutative group relative to the operation composition of functions.

   ii) The set of multiplication operations $\mathscr{M}$ is a commutative group relative to the operation composition.

3. The intersection of the sets $\mathscr{A}$ and $\mathscr{M}$ is not empty. There is exactly one function I which is in both these sets. $A_x = I = M_y$. What values of x and y determine I?

4. Compute:

   a) $A_2A_3(5)$ b) $M_1M_3(4)$ c) $A_{-4}A_4(7)$

   d) $M_{2/3}M_{3/2}(6)$ e) $A_0A_4(3)$ f) $M_{1/3}M_{1/4}(24)$

5. Solve each equation for x.

   a) $A_3(x) = 7$ b) $M_2(x) = 5$ c) $A_2A_3(x) = 10$

   d) $M_6M_{1/2}(x) = 12$ e) $A_{-4}A_6(x) = 2$ f) $M_{3/2}M_{2/3}(x) = 15$

6. Explain the solution of each equation shown below.

   a) If $A_2A_4(x) = 8$ b) If $M_{1/2}M_{2/3}(x) = 7$

   $\qquad A_6(x) = 8$ $\qquad M_{1/3}(x) = 7$

   $A_{-6}A_6(x) = A_{-6}(8)$ $M_3M_{1/3}(x) = M_3(7)$

   $\qquad A_0(x) = A_{-6}(8)$ $\qquad M_1(x) = M_3(7)$

   $\qquad\qquad x = 2$ $\qquad\qquad x = 21$

7. In the set $\mathscr{L}$ we have not only all the functions of $\mathscr{A}$ and $\mathscr{M}$ but also functions
   like A_2M_3, $M_{-2}A_4M_3$, ... Note that in $\mathscr{L}$ the binary operation, composition
   of functions, is *not* commutative, for $M_2A_3(x) = 2(x + 3)$ while $A_3M_2(x) =$
   $2x + 3$. Clearly, $M_2A_3 \neq A_3M_2$. The set $\mathscr{L}$ is a group relative to the operation composition, but it is not a commutative group. Determine x, y, z, so that

   a) $A_2M_2 = M_2A_x$ b) $A_xM_5 = M_5A_2$

   c) $A_2M_3A_4 = A_2A_xM_3 = A_yM_3 = M_3A_z$

   d) $M_2A_4M_3A_5 = A_xM_2M_3A_5 = A_xM_yA_5 = A_zM_y$

   e) If $r \neq 0$, $A_sM_r = M_rA_x$

8. The main significance of the sytem $\mathscr{L}$ for students of elementary arithmetic
   is illustrated by problems of the following type. Discuss the type of reasoning
   students might use in solving the indicated problems.

   a) A third-grade teacher drills her class by inviting them to play the following
   game. She will call out a number. Each student is then first to add 6 and then
   to add 4, and think of the final sum.

   b) Describe an analog of (a) for multiplication. At what grade level might it
   be appropriate?

   c) The teacher of a third-grade class announces that she will think of a
   number, add 4 to her number, and give the result. Students are to discover
   her chosen number.

   d) A teacher of the sixth grade announces that she will choose a secret number,
   add 5, multiply by 2, and then give the final result. Students are to find the
   secret number.

   e) In a sixth-grade class the teacher presents a trick. Each child writes down
   his age, multiplies by 5, adds 20, multiplies by 4, subtracts 43, multiplies by 5

and adds the age of his mother. The result is a four-digit number. From this result the teacher can compute both the student's age and the mother's age. How does she do it?

9. Solve each equation for x.

a) $M_2 A_6(x) = 20$
b) $M_{1/2} A_{-2}(x) = 4$
c) $A_6 M_2(x) = 20$
d) $A_{-2} M_{1/2}(x) = 4$
e) $A_3 M_2 A_{-4}(x) = 9$
f) $A_{-4} M_2 A_3(x) = 9$

10. Problem 9(a) may be written as $2(x + 6) = 20$. Write each of the other parts of Exercise 9 in this style.

11. Explain the solution shown for each equation below. Rewrite each equation in the notation of school algebra and solve.

a) If $\quad M_3 A_5(x) = 36,$
$$A_{-5} M_{1/3} M_3 A_5(x) = A_{-5} M_{1/3}(36)$$
$$x = 7$$

b) If $\quad A_{-4} M_{2/3} A_6(x) = 6,$
$$A_{-6} M_{3/2} A_4 A_{-4} M_{2/3} A_6(x) = A_{-6} M_{3/2} A_4(6)$$
$$x = 9$$

12. Each equation below describes a computational shortcut. Explain and illustrate by examples. (Here it is more natural to refer to A_{-1} as *subtract one*.)

a) $A_9 = A_{-1} A_{10}$
b) $M_{40} = M_4 M_{10}$
c) $A_{98} = A_{-2} A_{100}$
d) $A_{-998} = A_2 A_{-1000}$
e) $M_{50} = M_{1/2} M_{100}$
f) $M_{25} = M_{1/4} M_{100}$
g) $M_{125} = M_{1/8} M_{1000}$
h) $M_{375} = M_3 M_{1/8} M_{1000}$

13. Solve each equation for X.

a) $X A_5 M_3 = M_1 = I$
b) $A_5 M_3 X = I$
c) $X M_2 A_{-4} M_3 = A_0 = I$
d) $M_2 A_{-4} M_3 X = I$

14. For each part of Exercise 13 designate the inverse of the function X.

15. Determine r, s, and t so that $M_4 A_{12} M_3 A_4 M_2 A_1 = A_s M_r = M_r A_t$.

16. Solve the following problem: I spent one-third of my money, then spent \$2, spent one-fourth of the remainder, and then finally spent \$5. If I had \$7 left, how much had I originally?

17. From a basket of eggs were removed $\frac{1}{2}$ of an egg more than $\frac{1}{2}$ of all the eggs. This operation was repeated again and yet again (i.e., 3 times in all). There remained 3 eggs in the basket. How many were there at the beginning? Relate this problem to each of the equations below.

$$A_{-1/2} M_{1/2} A_{-1/2} M_{1/2} A_{-1/2} M_{1/2}(x) = 3; \qquad M_{1/2} A_{-1} M_{1/2} A_{-1} M_{1/2} A_{-1}(x) = 3$$

5–4 Geometrical Interpretations

The connection between geometry and algebra depends on the concept of a number line. We can visualize numbers as points on a line, and operations on numbers can be given geometric interpretations. In the elementary school, students deal with three sets of numbers and their number-line representations. These are the sets:

$$W = \{0, 1, 2, \ldots\}, \qquad I = \{\ldots, -2, -1, 0, 1, 2, \ldots\},$$
$$Q^+ = \{0, 1, 2, \ldots, \tfrac{1}{2}, \tfrac{1}{3}, \tfrac{1}{4}, \tfrac{2}{3}, \ldots\}.$$

The whole number line is a matching of the whole numbers with some of the points of a ray.

The integer line utilizes the other ray.

The number line for Q^+ assigns numbers to many points between the whole number points. When we assign numbers to points in this fashion we say that we are establishing a *coordinate system* on the line.

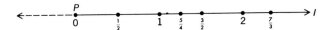

For these three number lines the geometrical interpretations of addition, subtraction, multiplication, and division are quite simple. Basically, all these operations are viewed as *counting* operations. With the introduction of the real numbers (i.e., the invention of irrational numbers), geometric interpretations become more complicated. It is easy to see where the difficulty lies. With nothing more than a compass you can "construct" points for as many integers as you please. And given any two numbers to add, you can explain how to position your compass, draw circles, and locate the point for the sum. In order to locate the points on a line which are matched with fractions, you need only a compass and straightedge (an unmarked ruler). And once these points are located, the operations of addition, multiplication, etc., can be described in terms of simple counting. But we have no way to locate the points on a line which are matched with many of our irrational numbers. We can find the point for $\sqrt{2}$ (How?), but we cannot locate the point for $\sqrt[3]{2}$, using our ruler and compass.

Given a particular real number, we cannot in general prescribe a finite number of geometric operations with ruler and compass that will enable us to locate the point for that number. And if we have points for two real numbers, like

$$\sqrt[3]{4} \quad \text{and} \quad \sqrt[5]{7},$$

there is no way of utilizing counting techniques in locating the point for their product.

The exercises below present geometric interpretations of number operations on these number lines.

Exercises 5–3

1. The integer number line is said to have *holes* in it because there are points of the line which are not matched with any integers. Prove that the integer line has holes in it. Prove also that the rational number line has holes.

2. Explain how to use a compass to add and subtract on the integer number line. What distinction would you make between computing geometrically the sums $2 + 3$ and $3 + 2$?

3. The diagrams below present standard geometrical interpretations of whole number multiplication and division as they are usually presented to elementary students. Explain these diagrams.

   a) What product and quotient are pictured?

   b) Sketch a diagram picturing: 2×3; $20 \div 4$; and $20 \div 3$ (showing quotient and remainder for the latter).

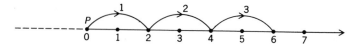

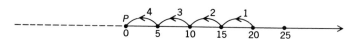

4. Explain the ruler-compass construction below for locating fifths on the rational number line.

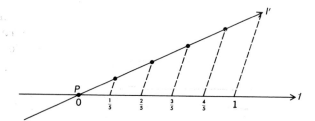

5. The diagram below pictures the product $\frac{2}{3} \times \frac{4}{7}$, where of course we think of this as the task of finding two-thirds of a segment of length $\frac{4}{7}$. Explain the diagram, sketch a similar figure for $\frac{4}{7} \times \frac{2}{3}$, and give a geometrical argument (which will involve some counting) to show that $\frac{2}{3} \times \frac{4}{7} = \frac{8}{21}$.

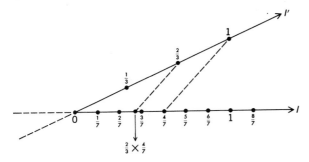

6. Geometric interpretations of real number addition and multiplication are presented below. These interpretations utilize properties of parallel lines and similar triangles. In the first figure, $l \parallel l'$. Draw corresponding diagrams for $r - s$ and $r \div s$.

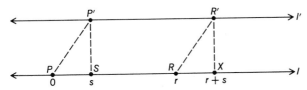

Addition of real numbers

7. As in Exercise 6, sketch diagrams depicting
   a) $r + r$ b) $r \cdot r$ c) $r \div r$
   d) $1 \div r$, (i) for $r > 1$ and (ii) for $r < 1$.
   e) $(-1) \cdot s$ f) $r \div (-1)$
   g) $s \cdot r$ and $s + r$, as contrasted with $r \cdot s$ and $r + s$.

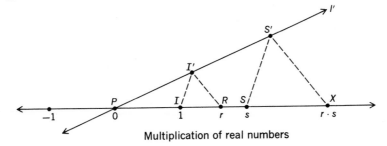

Multiplication of real numbers

8. The figure below presents a construction for the square root of any positive real number x. Construct the circle having points 0 and $x + 1$ as endpoints of a diameter. The length of $\overline{QX}$ is the square root of x.

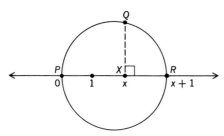

a) Show that triangles PQR, PXQ, and QXR are all similar to one another and argue that the length of $\overline{QX}$ is $\sqrt{x}$. An alternate proof would be to show that the area of a square with $\overline{QX}$ as a side is the same as the area of a rectangle whose sides have length 1 and x, respectively.

b) What happens when you attempt to apply the above construction to determine the square root of a negative number x?

If you remember graphing linear functions in school algebra, you have already one geometrical interpretation of the linear functions we considered in Section 5–3. Mathematical terminology usually introduced when this topic is studied includes *slope* and *intercept*. We shall not review these ideas in detail. The graphs below are presented to help you recall the concepts. Discuss these graphs.

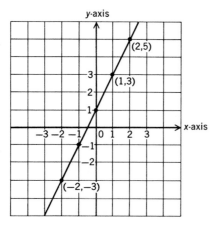

Graph of the linear function $2x + 1$. The slope is 2, the y- intercept is 1, and the x-intercept is $-1/2$.

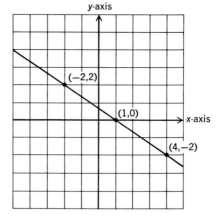

Give slope, intercepts, and name of the linear function pictured.

The graphical representation of linear functions already depicted fails to emphasize some of their characteristics that have significance for geometry. If we think of the real numbers as points on a line then we can interpret linear functions according to the sketches below.

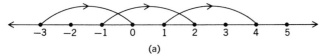

(a)

We may think of A_3 as an operation (function) which *translates* each point of the real number line 3 units in the "positive" direction.

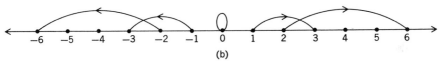

(b)

We may think of M_3 as a function which *stretches* the number line about the fixed point 0, moving each point to a position 3 times as far from the origin.

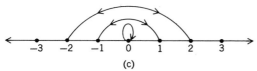

(c)

We may think of M_{-1} as a function which *reflects* the points of the line in the origin or rotates the line about 0 through a 180° angle. We call M_{-1} a *half turn* about 0.

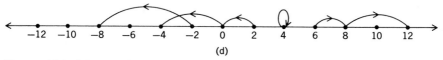

(d)

We may think of $A_{-4} M_2$ (or $2x-4$) as the *composition* of a stretching about the origin 0 and then a translation to the left. But note. We can think of $A_{-4} M_2$ as simply a stretching about point 4. Explain.

Exercises 5–4

1. Each linear function below can be interpreted as a stretching of the number line about some fixed point. Determine this fixed point and the stretching factor.

   a) $2x - 12$ b) $2x + 12$ c) $3x - 12$

   d) $4x + 12$ e) $5x - 12$ f) $6x + 11$

2. In the diagram for 1 (c), we saw that the linear function M_{-1} described by the equation $M_{-1}(x) = 0 - x$ can be visualized as a half-turn about the point 0. Each function below is also a half-turn about a center other than 0. Verify this assertion.

a) $2 - x$ b) $8 - x$ c) $-4 - x$

3. The example below illustrates how one may calculate the composite of two linear functions.

If $L_1(x) = 2x - 4$, $L_2(x) = -3x + 5$, then

$$L_1 = A_{-4}M_2, \qquad L_2 = A_5M_{-3} \qquad \text{and} \qquad L_1L_2 = A_{-4}M_2A_5M_{-3}.$$

Hence

$$L_1L_2(x) = 2(-3x + 5) + (-4) = -6x + 6.$$

a) For L_1, L_2 as above, compute $L_2L_1(x)$.

b) Compute $L_1L_2(x)$ and $L_2L_1(x)$, given that $L_1(x) = 8 - x$, $L_2(x) = 3x + 4$.

c) If $L_1(x) = 7x + 3$, $L_2(x) = 5x - 7$, how can the formulas

$$L_1L_2(x) = 7(5x - 7) + 3$$

and

$$L_2L_1(x) = 5(7x + 3) - 7$$

be written at once?

4. Let $H_1(x)$ and $H_2(x)$ be the half-turns described by $8 - x$ and $12 - x$, respectively. Then

$$H_1H_2(x) = x + 4$$

and

$$H_2H_1(x) = x - 4.$$

 *reverse results!*

That is, the composite of these two half-turns is a translation four units to the right or left depending upon the order of composition. Generalize this result and show that if H_1 and H_2 are any half-turns, then H_1H_2 and H_2H_1 are translations of equal distances in opposite directions.

5–5 Linear Fractional Transformations

Another class of algebraic functions which has geometric importance is the group of linear fractional transformations. These are functions described by fractional expressions of the type

$$\frac{ax + b}{cx + d}.$$

Note that both numerator and denominator are linear functions. These linear fractional functions are important in projective geometry. We shall not stress their geometric significance but will treat them algebraically. These transformations provide interesting examples of groups of functions which contain only a few elements.

We shall consider only functions $(ax + b)/(cx + d)$, where a, b, c, d are whole numbers. (If we permitted a, b, c, d to be rational numbers, we would not get any new functions. Why?) When we restrict the coefficients a, b, c, d in this way, we find that the only *finite* groups we obtain contain 1, 2, 3, 4, 6, 8, or 12 elements. All other groups have infinitely many elements. As an example, let

$$F_1(x) = \frac{x}{x - 1} \qquad \text{and} \qquad F_2(x) = \frac{x - 2}{2x - 1}.$$

Then $F_1^2 = F_1 F_1$ is the identity function I, as is also F_2^2. Note the calculation below which proves that $F_1^2 = I$. (Explain.)

$$F_1 F_1(x) = F_1\left(\frac{x}{x - 1}\right) = \frac{\dfrac{x}{x - 1}}{\dfrac{x}{x - 1} - 1} = x.$$

We say that F_1 has order 2. To attack the problem geometrically, we can begin with any point on the number line, apply F_1 twice and return to our starting point. Carry out the calculation showing that $F_2^2 = I$, also.

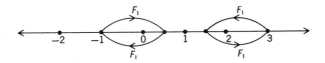

Now when we compute $F_1 F_2$ and $F_2 F_1$ we find that they are equal to each other.

$$F_1 F_2(x) = F_1\left(\frac{x - 2}{2x - 1}\right) = \frac{\dfrac{x - 2}{2x - 1}}{\dfrac{x - 2}{2x - 1} - 1} = \frac{x - 2}{-x - 1}.$$

Compute $F_2 F_1$ and show that $F_2 F_1 = F_1 F_2$.

With this knowledge we can now show that with F_1 and F_2, the only group we can build is one of 4 elements. For

$$F_1 F_2 F_1 = F_2 F_1 F_1 = F_2 I = F_2. \quad \text{(Explain.)}$$

The table of composition for the group generated by F_1 and F_2 is given below. Complete this table.

·	I	F_1	F_2	F_1F_2
I	I	F_1	F_2	F_1F_2
F_1	F_1	I		
F_2	F_2		I	
F_1F_2	F_1F_2			I

The fact that F_1 and F_2 generate this group of order 4 has a geometric interpretation. If we choose a point on the number line and successively apply the operations F_1 and F_2, we can never reach more than 4 different points.

The function

$$F: x \to \frac{1}{1 - x}$$

has order 3. That is, applying F three times to any point returns us to that point.

$$F: 2 \to -1 \to \tfrac{1}{2} \to 2, \qquad F: \tfrac{3}{4} \to 4 \to -\tfrac{1}{3} \to \tfrac{3}{4},$$

$$F: a \to \frac{1}{1-a} \to \frac{1}{1 - \dfrac{1}{1-a}} = \frac{1-a}{-a} \to \frac{1}{1 - \dfrac{1-a}{-a}} = a.$$

Check the last calculation above and explain why this is a proof that $F^3 = I$.

This function F generates a group of 3 elements. Explain the table below.

·	I	F	F^2
I	I	F	F^2
F	F	F^2	I
F^2	F^2	I	F

If we combine with F the function

$$G: x \to \frac{1}{x},$$

we find that these two functions generate a group of 6 elements. The table for this group is begun below. Complete this table. Note that the group is not commutative; that $G^2 = I$, and $F^2G = GF$. (Verify these assertions.)

$\cdot$	I	F	F^2	G	FG	GF
I	I	F	F^2	G	FG	GF
F	F	F^2	I	FG	GF	
F^2	F^2	I	F	GF		
G	G	GF				
FG	FG					
GF	GF					

We should note one unusual property of these new functions. The function F of order 2 with $F(x) = 1/x$ does not operate upon the number 0. Moreover, no matter how large x may be, $1/x$ is never 0. But if x is very large, $1/x$ is near 0, and if x is near 0, $1/x$ is very large. It is convenient to invent a new point on our number line that we call the *point at infinity*. To this point we assign the symbol ∞ (call it infinity, if you please). We agree that $F(0) = \infty$ and $F(\infty) = 0$. With this agreement the function F is "nicer" than before. It operates on every point of our line. This invention of "points at infinity" is an essential characteristic of projective geometry, the geometry that makes use of linear fractional transformations.

For the function

$$G: x \to \frac{x - 1}{2x - 1}$$

verify that for x near $\frac{1}{2}$, $G(x)$ is numerically very large; and when x is very large, $G(x)$ is near $\frac{1}{2}$. Hence we agree that $G(\frac{1}{2}) = \infty$ and $G(\infty) = \frac{1}{2}$. The function

$$H: x \to \frac{1}{1 - x}$$

has order 3. We agree that $H(1) = \infty$ (Why?); $H(\infty) = 0$ (Why?); and now note that $H(0) = 1$. Hence,

$$H: 1 \to \infty \to 0 \to 1.$$

The function

$$R:\ x \to \frac{2x-1}{4x+2}$$

has order 4. Justify the calculations below.

$$R: 1 \to \tfrac{1}{6} \to -\tfrac{1}{4} \to -\tfrac{3}{2} \to 1.$$
$$R: 0 \to -\tfrac{1}{2} \to \infty \to \tfrac{1}{2} \to 0.$$
$$R: -1 \to \tfrac{3}{2} \to \tfrac{1}{4} \to -\tfrac{1}{6} \to -1.$$

It is convenient to invent new symbols called *matrices* as names for these functions. We name the function

$$\frac{ax+b}{cx+d}$$

by the matrix symbol

$$\begin{pmatrix} a & b \\ c & d \end{pmatrix}.$$

The reason for inventing these symbols is that they lend themselves nicely to computation. For example, if

$$F_1:\ x \to \frac{x+2}{3x+4}, \qquad F_2:\ x \to \frac{x-1}{-2x+5},$$

then

$$F_1 = \begin{pmatrix} 1 & 2 \\ 3 & 4 \end{pmatrix}, \qquad F_2 = \begin{pmatrix} 1 & -1 \\ -2 & 5 \end{pmatrix},$$

and we can compute the matrix name for the composite function $F_1 F_2$ by a "row by column" technique for multiplying matrices. The formula for this matrix multiplication is

$$\begin{pmatrix} a & b \\ c & d \end{pmatrix}\begin{pmatrix} A & B \\ C & D \end{pmatrix} = \begin{pmatrix} aA+bC & aB+bD \\ cA+dC & cB+dD \end{pmatrix}.$$

Note that the number $aA + bC$ which occurs in the *first row* and *first column* of the product matrix is computed by combining the first row $(a \ \ b)$ of the matrix on the left with the first column $\begin{pmatrix} A \\ C \end{pmatrix}$ of the right-hand matrix as shown below.

$$(a \quad b) \quad \begin{pmatrix} A \\ C \end{pmatrix} \rightarrow aA + bC.$$

The number $cA + dC$ in the second row, first column of the product, arises in the same way from the second row $(c\ d)$ and first column $\begin{pmatrix} A \\ C \end{pmatrix}$. Verify now that

$$F_1 F_2 = \begin{pmatrix} 1 & 2 \\ 3 & 4 \end{pmatrix}\begin{pmatrix} 1 & -1 \\ -2 & 5 \end{pmatrix} = \begin{pmatrix} -3 & 9 \\ -5 & 17 \end{pmatrix} = \frac{-3x + 9}{-5x + 17},$$

$$F_2 F_1 = \begin{pmatrix} 1 & -1 \\ -2 & 5 \end{pmatrix}\begin{pmatrix} 1 & 2 \\ 3 & 4 \end{pmatrix} = \begin{pmatrix} -2 & -2 \\ 13 & 16 \end{pmatrix} = \frac{-2x - 2}{13x + 16}.$$

The exercises below continue the development of the ideas that have been introduced in this section.

Exercises 5–5

1. The identity function I with

$$I(x) = x = \frac{x}{1}$$

has matrix name

$$\begin{pmatrix} 1 & 0 \\ 0 & 1 \end{pmatrix}.$$

Verify that for any matrix

$$\begin{pmatrix} a & b \\ c & d \end{pmatrix},$$

$$\begin{pmatrix} a & b \\ c & d \end{pmatrix}\begin{pmatrix} 1 & 0 \\ 0 & 1 \end{pmatrix} = \begin{pmatrix} 1 & 0 \\ 0 & 1 \end{pmatrix}\begin{pmatrix} a & b \\ c & d \end{pmatrix} = \begin{pmatrix} a & b \\ c & d \end{pmatrix}.$$

2. Actually every linear fractional function has infinitely many matrix names (just as every rational number has infinitely many fraction names).

   a) Show that each of

   $$\begin{pmatrix} -1 & 0 \\ 0 & -1 \end{pmatrix}, \quad \begin{pmatrix} 3 & 0 \\ 0 & 3 \end{pmatrix}, \quad \begin{pmatrix} a & 0 \\ 0 & a \end{pmatrix}$$

   is a name for the identity function I.

b) Verify that

$$\begin{pmatrix} 6 & 12 \\ 15 & -3 \end{pmatrix} = \begin{pmatrix} 4 & 8 \\ 10 & -2 \end{pmatrix},$$

that is, these two matrices name the same function. Find the "simplest" matrix name for this function.

3. Write a matrix name for each function below.

a) $F(x) = \dfrac{x + 2}{3x - 1}$ b) $F(x) = \dfrac{3 - x}{4x}$ c) $F(x) = \dfrac{5 - 2x}{3}$

d) $F(x) = 3x - 2$ e) $F(x) = 8 - x$ f) $F(x) = \dfrac{1}{x}$

4. Let

$$F = \begin{pmatrix} -1 & 0 \\ 0 & 1 \end{pmatrix}, \qquad G = \begin{pmatrix} 1 & 1 \\ -1 & 1 \end{pmatrix},$$

that is

$$F: x \to -x \qquad \text{and} \qquad G: x \to \frac{x + 1}{-x + 1}.$$

Show that

a) $F^2 = I$ b) $G^4 = I$ c) $FG \neq GF$ d) $FG = G^3 F$

Now using just the facts (a), (b), (c), (d) above, complete the table begun below and show that F and G generate a group of 8 elements.

$\cdot$	I	F	G	G^2	G^3	GF	G^2F	G^3F
I	I	F	G	G^2	G^3	GF	G^2F	G^3F
F	F	I	G^3F					
G	G	GF	G^2					
G^2	G^2	G^2F						
G^3	G^3							
GF	GF							
G^2F	G^2F							
G^3F	G^3F							

5. A geometric interpretation of Exercise 4 is that by starting at one point and repeatedly applying functions F and G, one can reach no more than 8 points on the line.

   a) Show that beginning with $\frac{1}{2}$ one can reach the 8 points

   $$\tfrac{1}{2},\ -\tfrac{1}{2},\ 3,\ -3,\ \tfrac{1}{3},\ -\tfrac{1}{3},\ 2,\ -2.$$

   b) What points can be reached beginning with $\frac{1}{4}$? With 0? With 10?

6. Show that the functions

   $$F\colon x \to -x - 2 \qquad \text{and} \qquad G\colon x \to \frac{4}{3x}$$

   generate a group of order 12. [*Hint:* Show that $F^2 = G^2 = I$; $FG \neq GF$; $(FG)^2 \neq I$; $(FG)^3 \neq I$; $(FG)^6 = I$; $FG = (GF)^{-1}$; $(FG)^3 = (GF)^3$.]

7. Note that

   $$F = \begin{pmatrix} 2 & 3 \\ 4 & 6 \end{pmatrix}$$

   sends every point except $x = -\frac{3}{2}$ to the point $\frac{1}{2}$. Show that if

   $$F = \begin{pmatrix} a & b \\ c & d \end{pmatrix} \qquad \text{and} \qquad ad - bc = 0,$$

   then if

   $$x \neq -\frac{d}{c} \qquad \text{and} \qquad c \neq 0,$$

   we have

   $$f(x) = \frac{a}{c}.$$

8. Every function of order 2 is given by the formula

   $$F = \begin{pmatrix} a & b \\ c & -a \end{pmatrix}.$$

   Substitute various values for a, b, c and verify that in each case $F^2 = I$ unless $a^2 + bc = 0$.

9. Every function of order 3 is given by the formula

   $$F(x) = \frac{abx + b^2}{-(a^2 + ac + c^2)x + bc}.$$

Assign several sets of values to a, b, and c, and show that each gives a function such that $F^3 = I$.

10. Using the formula of Exercise 9, find values of a, b, c such that

$$F = \frac{x + 3}{-7x + 4}.$$

Verify that F has order 3.

11. No function with integer coefficients has order 5. However, verify that

$$\text{if} \quad F: x \to \frac{x - (5 + 2\sqrt{5})}{x + 1}, \quad \text{then} \quad F^5 = I.$$

12. For the matrix

$$\begin{pmatrix} a & b \\ c & d \end{pmatrix}$$

we call the number $ad - bc$ the *determinant* of the matrix and the number $a + d$ the *trace* of the matrix. If

$$F = \frac{ax + b}{cx + d}, \quad \text{and} \quad ad - bc \neq 0,$$

then

a) If $(a + d)^2 = 0 \cdot (ad - bc)$, F has order 2.

b) If $(a + d)^2 = 1 \cdot (ad - bc)$, F has order 3.

c) If $(a + d)^2 = 2 \cdot (ad - bc)$, F has order 4.

d) If $(a + d)^2 = 3 \cdot (ad - bc)$, F has order 6.

Construct a matrix of each type and check the above assertions.

13. Show that if $(RS)^3 = I$ for matrices R and S, then also $(SR)^3 = I$.

14. If we denote by R the function which matches each number with its reciprocal (i.e., $R: x \to 1/x$), then each linear fractional function is the composite of linear functions and R. For example,

i) If $F(x) = \dfrac{1}{x + 1}$, $F = RA_1$,

$$\left[\text{where } A_1 = x + 1 = \begin{pmatrix} 1 & 1 \\ 0 & 1 \end{pmatrix} \right].$$

ii) If $F(x) = \dfrac{2x + 7}{x + 2}$, $F = A_2 M_3 R A_2$,

$$\left[\text{where } M_3 = 3x = \begin{pmatrix} 3 & 0 \\ 0 & 1 \end{pmatrix} \right].$$

a) Verify (i) and (ii) above. Show that

$$A_{-1}R = \begin{pmatrix} 1 & -1 \\ 0 & 1 \end{pmatrix}\begin{pmatrix} 0 & 1 \\ 1 & 0 \end{pmatrix} = \begin{pmatrix} -1 & 1 \\ 1 & 0 \end{pmatrix} = \frac{-x+1}{x}$$

is the inverse of RA_1, and compute the inverse of

$$\frac{2x+7}{x+2}.$$

b) Factor

$$\frac{2x+3}{x-2}$$

as above into a product of addition functions A_x, multiplication functions M_y, and the "reciprocation" function R. Use this factorization of

$$\frac{2x+3}{x-2}$$

to construct its inverse. Using matrix notation and using the fact that

$$A_x = \begin{pmatrix} 1 & x \\ 0 & 1 \end{pmatrix}, \qquad M_y = \begin{pmatrix} y & 0 \\ 0 & 1 \end{pmatrix}, \qquad R = \begin{pmatrix} 0 & 1 \\ 1 & 0 \end{pmatrix},$$

write the matrix

$$\begin{pmatrix} 2 & 3 \\ 1 & -2 \end{pmatrix}$$

as a product of these simple matrices.

c) Try to write

$$\frac{ax+b}{cx+d}$$

as the composite of simple matrices A_x, M_y, and R.

15. Show that the matrix

$$\begin{pmatrix} 1 & 1 \\ 0 & 1 \end{pmatrix}$$

generates a group of infinitely many elements.

5–6 Groups of Permutations

Algebraic systems known as permutation groups have applicability to the theory of symmetric figures in geometry. Here is a simple illustrative example. We say that the rectangle below has four *symmetries*: the identity function (which leaves each point fixed), the reflections in l and m, and the half-turn about P.

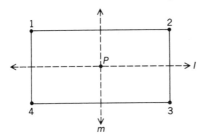

Each of these symmetries is characterized by the way it permutes the vertices of the rectangle. For example, if we flop the rectangle over on line l, then vertices in positions 1 and 4 exchange places, as do those in 2 and 3. We use the following symbols to name the four symmetries of the rectangle:

$$\begin{pmatrix} 1 & 2 & 3 & 4 \\ 1 & 2 & 3 & 4 \end{pmatrix}, \qquad\qquad \begin{pmatrix} 1 & 2 & 3 & 4 \\ 4 & 3 & 2 & 1 \end{pmatrix},$$

$$\text{The identity} \qquad\qquad\qquad \text{Reflection in } l$$

$$\begin{pmatrix} 1 & 2 & 3 & 4 \\ 2 & 1 & 4 & 3 \end{pmatrix}, \qquad\qquad \begin{pmatrix} 1 & 2 & 3 & 4 \\ 3 & 4 & 1 & 2 \end{pmatrix}.$$

$$\text{Reflection in } m \qquad\qquad \text{180}° \text{ Rotation} \\ \text{about } P$$

These symbols are self-explanatory.

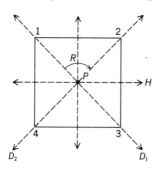

For the square above, we denote by

$$\begin{pmatrix} 1 & 2 & 3 & 4 \\ 2 & 3 & 4 & 1 \end{pmatrix}$$

the 90° clockwise rotation that moves the vertex in position 1 to position 2, that in position 2 to position 3, etc. The symbol

$$\begin{pmatrix} 1 & 2 & 3 & 4 \\ 1 & 4 & 3 & 2 \end{pmatrix}$$

represents the reflection in the diagonal D_1.

We can use these symbols to compute the composite of symmetries. For example, if we first rotate $90°$ clockwise around P and then reflect in the diagonal D_1, the final result is the same as if we had simply reflected in H.

$$D_1 R = \begin{pmatrix} 1 & \underline{2} & 3 & 4 \\ 1 & \underline{4} & 3 & 2 \end{pmatrix}\begin{pmatrix} \underline{1} & 2 & 3 & 4 \\ \underline{2} & 3 & 4 & 1 \end{pmatrix} = \begin{pmatrix} 1 & 2 & 3 & 4 \\ 4 & 3 & 2 & 1 \end{pmatrix} = H.$$

The underlined entries on the left indicate how we compute that vertex 1 is moved to the 4 position. We see that $1 \to 2 \to 4$; similarly, $2 \to 3 \to 3$. Verify that 3 and 4 are carried to positions 2 and 1, respectively. Of course we can use our geometric intuition and see that first rotating and then reflecting in D_1 has the same total effect as the reflection in H, but this symbolism and the accompanying computational technique enables us to get the result mechanically.

In the exercises below we shall view groups of symmetries for several geometric figures as permutations of the vertices.

Exercises 5–6

1. The line segment pictured has two symmetries, the identity

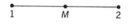

$$\begin{pmatrix} 1 & 2 \\ 1 & 2 \end{pmatrix}$$

and the half-turn about M,

$$\begin{pmatrix} 1 & 2 \\ 2 & 1 \end{pmatrix}.$$

Denote

$$\begin{pmatrix} 1 & 2 \\ 1 & 2 \end{pmatrix} \text{ by } I \quad \text{ and } \quad \begin{pmatrix} 1 & 2 \\ 2 & 1 \end{pmatrix} \text{ by } R,$$

and then explain the group table.

·	I	R
I	I	R
R	R	I

2. Explain and complete the group table for the rectangle shown below. Is this a commutative group?

$$I = \begin{pmatrix} 1 & 2 & 3 & 4 \\ 1 & 2 & 3 & 4 \end{pmatrix}, \qquad H = \begin{pmatrix} 1 & 2 & 3 & 4 \\ 4 & 3 & 2 & 1 \end{pmatrix},$$

$$V = \begin{pmatrix} 1 & 2 & 3 & 4 \\ 2 & 1 & 4 & 3 \end{pmatrix}, \qquad R = \begin{pmatrix} 1 & 2 & 3 & 4 \\ 3 & 4 & 1 & 2 \end{pmatrix}.$$

·	I	H	V	R
I	I	H	V	R
H	H	I		
V	V		I	
R	R			I

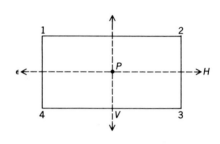

3. We denote the symmetries of the equilateral triangle in the accompanying figure by:

$$I = \begin{pmatrix} 1 & 2 & 3 \\ 1 & 2 & 3 \end{pmatrix}, \qquad a = \begin{pmatrix} 1 & 2 & 3 \\ 2 & 1 & 3 \end{pmatrix}, \qquad b = \begin{pmatrix} 1 & 2 & 3 \\ 3 & 2 & 1 \end{pmatrix},$$

$$c = \begin{pmatrix} 1 & 2 & 3 \\ 1 & 3 & 2 \end{pmatrix}, \qquad R = \begin{pmatrix} 1 & 2 & 3 \\ 2 & 3 & 1 \end{pmatrix}, \qquad R^2 = \begin{pmatrix} 1 & 2 & 3 \\ 3 & 1 & 2 \end{pmatrix}.$$

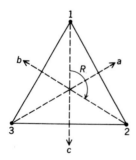

Interpret each symmetry geometrically as a rotation or reflection, and complete the table. Is the group commutative?

·	I	R	R^2	a	b	c
I	I	R	R^2	a	b	c
R	R	R^2	I	b		
R^2	R^2	I	R			
a	a					
b	b					
c	c					

4. As in Exercise 3, list the symmetries of a square and construct the table of composition for the group of 8 elements.

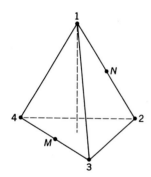

5. For the regular tetrahedron shown, interpret

$$\begin{pmatrix} 1 & 2 & 3 & 4 \\ 1 & 3 & 4 & 2 \end{pmatrix}$$

as a 120° rotation about the altitude from vertex 1. Interpret

$$\begin{pmatrix} 1 & 2 & 3 & 4 \\ 2 & 1 & 4 & 3 \end{pmatrix}$$

as a reflection in the line MN, where M, N are midpoints of opposite sides. Show that the group of symmetries of this regular tetrahedron has 12 elements. Write symbols for all these permutations of the vertices.

6. As in Exercise 5 for the tetrahedron, consider the groups of symmetries for cube, octahedron, dodecahedron, and icosahedron.

6 Algebraic Structures II

6–1 Finite Rings and Fields

Many elementary arithmetic textbooks introduce modular arithmetic (sometimes called clock arithmetic). The clock motivates equations of the following type:

$11 + 3 = 2$ (3 hours after 11 o'clock it is 2 o'clock),

$1 - 3 = 10$ (3 hours before 1 o'clock it is 10 o'clock).

One can choose any whole number n larger than 1 and compute *modulo n*. For example, if $n = 6$, one adds or multiplies any two whole numbers smaller than 6 in this modular arithmetic by adding or multiplying in the usual way and then computing the remainder when this result is divided by 6. For example,

$4 + 5 = 3,$ since $9 - 6 = 3.$

$4 \cdot 5 = 2,$ since $20 - 18$ (three 6's) $= 2.$

In mod-6 arithmetic, one deals with only the six numbers,

0, 1, 2, 3, 4, 5.

Addition and multiplication tables in this finite arithmetic are given below.

+	0	1	2	3	4	5
0	0	1	2	3	4	5
1	1	2	3	4	5	0
2	2	3	4	5	0	1
3	3	4	5	0	1	2
4	4	5	0	1	2	3
5	5	0	1	2	3	4

·	0	1	2	3	4	5
0	0	0	0	0	0	0
1	0	1	2	3	4	5
2	0	2	4	0	2	4
3	0	3	0	3	0	3
4	0	4	2	0	4	2
5	0	5	4	3	2	1

In modular arithmetic both addition and multiplication are commutative and associative operations; moreover, multiplication is distributive over addition. The operation + has identity element 0, and the operation · has 1 for identity element. In mod-6 arithmetic each element has an additive inverse.

$$0 + 0 = 0; \qquad 1 + 5 = 0; \qquad 2 + 4 = 0; \qquad 3 + 3 = 0.$$

However, in mod-6 arithmetic there are nonzero elements that fail to have reciprocals (multiplicative inverses). Indeed, only the elements 1 and 5 have inverses. How does the multiplication table above show this?

Mod-6 arithmetic is an algebraic system that we call a *ring*. It fails to be a field because some of its nonzero elements lack multiplicative inverses. On the other hand, if we look at mod-5 arithmetic we find that all nonzero elements in this system do have reciprocals. Hence this system is a finite field of 5 elements. Explain the table below.

+	0	1	2	3	4
0	0	1	2	3	4
1	1	2	3	4	0
2	2	3	4	0	1
3	3	4	0	1	2
4	4	0	1	2	3

·	0	1	2	3	4
0	0	0	0	0	0
1	0	1	2	3	4
2	0	2	4	1	3
3	0	3	1	4	2
4	0	4	3	2	1

What makes a finite field like this mod-5 system so attractive is that school algebra is principally a study of properties of fields. Consequently, most of the important concepts of algebra can be illustrated by these simple systems. The exercises below develop these ideas.

Exercises 6–1

1. Compute in mod-6 arithmetic:
   - a) $(3 + 4) + 2$
   - b) $3 + (4 + 2)$
   - c) $5 + (1 + 3)$
   - d) $(5 + 1) + 3$
   - e) $3 \cdot (2 \cdot 5)$
   - f) $(3 \cdot 2) \cdot 5$
   - g) $4 \cdot (4 \cdot 4)$
   - h) $(4 \cdot 4) \cdot 4$

2. Give the negative of each number in mod-6 arithmetic.
   - a) 3   b) 5   c) 2   d) 0

3. Solve each equation below in mod-6 and classify each solution as
   - i) unique (eactly one number in the solution set),
   - ii) distinct (at least two different numbers in the solution set), or

iii) nonexistent (no number in the solution set).

a) $x + 3 = 2$ [Add the negative of 3 to each side.]

b) $5x = 2$ [Multiply each side by the reciprocal of 5.]

c) $5x + 4 = 1$ [Add the negative of 4 and multiply by the reciprocal of 5.]

d) $5(x + 3) = 4$ e) $2x = 3$

f) $3x = 5$ g) $2x = 0$

h) $2x = 4$ i) $3x + 2 = 5$

j) $x^2 = 4$ k) $x^2 = 3$

l) $(x + 3)^2 = 1$ m) $x^2 = 3$

n) $x^2 + 3x + 2 = 0$ o) $2x^2 + 3x + 5 = 0$

p) $x^3 = 1$ q) $x^3 = x$

r) $(x + 1)(x + 2)(x + 3) = 0$ s) $(x + 2)(x + 3)(x + 4) = 0$

t) $(x + 3)(x + 4)(x + 5) = 0$ u) $x^{1000} = x$

v) $x^{1001} = x$

4. In any field, if $a \cdot x = a \cdot c$ and $a \neq 0$, then multiplying by the reciprocal of a, one gets the result that $x = c$. However, in our mod-6 ring, certain nonzero elements fail to have reciprocals, so the conclusion that $x = c$ is not warranted. Each equation below has more than one solution. Determine these solutions.

a) $3 \cdot x = 3 \cdot 1$ b) $4 \cdot x = 4 \cdot 2$

c) $2 \cdot x = 2 \cdot 5$ d) $3 \cdot x = 3 \cdot 0$

e) $4 \cdot (x + 1) = 4 \cdot 0$ f) $3 \cdot (3x + 5) = 3 \cdot 2$

5. Verify that each factorization below is correct in the mod-6 system by multiplying the factors together. Note that one polynomial may factor in more than one way.

a) $x^2 + 3x + 2 = (x + 1)(x + 2)$

b) $x^2 + 3x + 2 = (x + 5)(x + 4)$

c) $x^2 + 5x = (x + 2)(x + 3)$

d) $x^2 + 5x = x(x + 5)$

e) $x^2 + 2 = (x + 2)(x + 4)$ [Can you factor $x^2 + 2$ in any other way?]

f) $x^2 + x = x(x + 1)$ [Can you factor $x^2 + x$ in any other way?]

g) $x^3 + 3x^2 + 2x = (x + 2)(x + 3)(x + 4)$

h) $x^3 + 3x^2 + 2x = x(x + 1)(x + 2)$

i) $x^3 + 3x^2 + 2x = x(x + 4)(x + 5)$

   j) $x^3 + 5x = (x + 1)(x + 2)(x + 3)$

N k) $x^3 + 5x = (x + 2)(x + 3)(x + 4)$ [Can you factor $x^3 + 5x$ in any other way?]

6. This, and all subsequent problems in the list are in the mod-5 field. Hence all computation will be reminiscent of school algebra. Compute, mod-5.

   a) $(3 + 4) + 1$ b) $3 + (4 + 1)$

   c) $2 \cdot (3 \cdot 4)$ d) $(2 \cdot 3) \cdot 4$

7. Give the negative of each number in the mod-5 system.

   a) 3   b) 2   c) 1   d) 0

8. Give the reciprocal of each number in the mod-5 system.

   a) 1   b) 2   c) 3   d) 4

9. Show that each linear equation below has a unique solution in the mod-5 system. Solve by adding negatives and multiplying by reciprocals.

   a) $x + 3 = 2$ b) $2x = 3$ c) $3x = 2$

   d) $2x + 4 = 1$ e) $4x + 3 = 2$ f) $3 \cdot (4x + 2) = 4$

10. Each quadratic equation below has no root, one root, or two distinct roots in mod-5. Solve and classify the equations.

   a) $x^2 = 1$ b) $x^2 = 2$

   c) $x^2 = 3$ d) $x^2 = 4$

   e) $x^2 = 0$ f) $x^2 + 3x + 2 = 0$

   g) $x^2 + 2x + 1 = 0$ h) $x^2 + 4x = 0$

   i) $x^2 + 1 = 0$ j) $x^2 + x + 3 = 0$

   k) $x^2 + 2x + 2 = 0$ l) $x^2 + x + 4 = 0$

   m) $x^2 + x + 1 = 0$ n) $x^2 + x + 2 = 0$

   o) $4x^2 + 4x + 3 = 0$

11. Prove that if the product of two numbers is zero, at least one of the numbers is zero. (A proof can be made by referring to the mod-5 table, but this is rather inelegant.)

12. Use the result of Exercise 11 to give all roots of the equations below.

   a) $(x + 2)(x + 3) = 0$ b) $x(2x + 5) = 0$

   c) $(2x + 1)(3x + 2)(x + 4) = 0$ d) $(x + 3)^2 = 0$

   e) $(x + 3)^2(3x + 1)^3 = 0$ f) $x^2(4x + 1)^3(4x + 3)^2 = 0$

13. The polynomial $x^2 + 1$ factors into $(x + 2)(x + 3)$. Verify this fact and also verify the other factorizations below.

   a) $x^2 + 1 = (2x + 4)(3x + 4)$

   b) $x^2 + 1 = (3x + 1)(2x + 1)$

   c) $x^2 + 1 = (4x + 3)(4x + 2)$

14. The factorizations (a), (b), (c) of $x^2 + 1$ in Exercise 13 are trivial variants of $(x + 2)(x + 3)$. For example, we can obtain (a) as shown below.

$$x^2 + 1 = (x + 2)(x + 3) = (2 \cdot 3)(x + 2)(x + 3) \quad [\text{Why?}]$$
$$= 2(x + 2) \cdot 3(x + 3)$$
$$= (2x + 4)(3x + 4).$$

   Obtain the trivial factorizations (b) and (c) similarly.

15. Factorizations for polynomials are given below. For each polynomial give at least two other factorizations which are trivial variations of the given one.

   a) $x^2 + 3x + 2 = (3x + 3)(2x + 4)$

   b) $2x^2 + x + 2 = (2x + 3)(x + 4)$

   c) $3x^2 + 3x + 2 = (3x + 4)(x + 3)$

16. For each equation below determine a, b, and c.

   a) $2x^2 + 3x + 1 = 2(ax^2 + bx + c)$

   b) $3x^2 + 2x + 4 = 3(ax^2 + bx + c)$

   c) $x^2 + 3x + 2 = 4(ax^2 + bx + c)$

   d) $3x^2 + x + 4 = 4(ax^2 + bx + c)$

17. Verify that the polynomial $x^5 + 4x$ factors into

$$x(x + 1)(x + 2)(x + 3)(x + 4).$$

18. Explain each step in the solution of the following quadratic equation. At what point is the result of Exercise 11 used?

$$4x^2 + x + 2 = 0$$

   1) $x^2 + 4x + 3 = 0$

   2) $(x + 1)(x + 3) = 0$

   3) $x = 4$ or $x = 2$

19. Solve each quadratic equation below by following the procedure used in Exercise 18.

   a) $3x^2 + 4x + 1 = 0$ b) $4x^2 + 3x = 0$

   c) $2x^2 + 3x + 3 = 0$ d) $3x^2 + x + 1 = 0$

20. Divide the polynomial $x^3 + 2x^2 + 4x + 1$ by $x + 1$, and show that

$$x^3 + 2x^2 + 4x + 1 = (x + 1)(x^2 + x + 3) + 3.$$

21. Use the result of Exercise 20 to prove that 4 is not a root of the equation $x^3 + 2x^2 + 4x + 1 = 0$.

We can introduce subtraction and division operations in the mod-5 system if we wish. We define

$$a - b = a + (-b)$$

$$\frac{a}{b} = a \div b = a \cdot b^{-1}$$

We also write

$$b^{-1} \quad \text{as} \quad \frac{1}{b}, \quad \text{so} \quad \frac{a}{b} = a \cdot \frac{1}{b}.$$

With these definitions we find that $(a - b) + b = a$ and $(a \div b) \cdot b = a$. (Show that this is so.) Of course these operations are not essential. For example, there is no need to use fraction notation, since

$$\tfrac{1}{2} = 3, \quad \tfrac{1}{3} = 2, \quad \tfrac{1}{4} = 4.$$

Each subtraction problem is easily converted into an addition problem. For example, $3 - 1 = 3 + (-1) = 3 + 4 = 2$, and $2 - 4 = 2 + (-4) = 2 + 1 = 3$. Introducing symbols like $\tfrac{2}{3}$ and -4 does not introduce new numbers into our mod-5 system. However, it is interesting to use these new symbols in mod-5 computation. The exercises below provide opportunities to do this.

Exercises 6–2

1. Perform each calculation below in mod-5 arithmetic, using the familiar rules for computing with fractions. Then verify each result by replacing all fractions by elements of the mod-5 system and recomputing.

   a) $\tfrac{1}{2} + \tfrac{1}{2}$ b) $\tfrac{1}{4} + \tfrac{1}{4}$ c) $1 + \tfrac{1}{3}$

   d) $\tfrac{2}{3} + \tfrac{2}{3} + \tfrac{2}{3}$ e) $\tfrac{3}{2} + \tfrac{2}{4}$ f) $\tfrac{1}{4} + (\tfrac{1}{2} + \tfrac{1}{4})$

2. Work each subtraction problem below by writing $a - b$ as $a + (-b)$.

   a) $4 - 1$ b) $3 - 2$ c) $4 - 3$

   d) $2 - 4$ e) $1 - 3$ f) $3 - 4$

3. Try to devise a proof that for all nonzero a, b in our mod-5 system:

   a) $\dfrac{1}{a} + \dfrac{1}{b} = \dfrac{b + a}{ab}$,   b) $\dfrac{1}{a} \cdot \dfrac{1}{b} = \dfrac{1}{ab}$.

Since there are only four choices for a and four for b, these proofs can be made by considering all possible cases. But this is a clumsy, unimaginative method of proof. Try to improve upon this.

4. Devise proofs of the following for all a, b, c in the mod-5 system:

   a) $(-a) + (-b) = -(a + b)$

   b) $a - (b - c) = a - b + c$

   c) $a(-b) = -(ab)$

   d) $(-a)(-b) = ab$

   e) $0 - a = -a$

5. Show that

   a) if $\dfrac{a}{b} = \dfrac{c}{d}$,   then $ad = bc$,

   b) if $ad = bc$, $b \neq 0$, and $d \neq 0$,   then $\dfrac{a}{b} = \dfrac{c}{d}$.

6. Show that

   a) $\dfrac{-a}{b} = \dfrac{a}{-b} = -\dfrac{a}{b}$,   b) $\dfrac{0}{a} = 0$,   c) $\dfrac{a}{1} = a$.

6–2 Algebras of Functions

In Chapter 5 we considered sets of functions which are groups under the operation *composition*. In school algebra two other operations upon functions are introduced, operations called *addition* and *multiplication*. You are familiar with these operations illustrated below for polynomial functions.

$$(x + 1) + (x + 2) = 2x + 3,$$

$$(x + 1) \cdot (x + 2) = x^2 + 3x + 2.$$

Perhaps you have not interpreted polynomials like

$$x + 1,\ x + 2,\ 2x + 3,\ x^2 + 3x + 2$$

as functions. However, when numbers are substituted for x in any polynomial a correspondence (function) between sets of numbers is obtained.

To fix the ideas, we consider polynomials with rational coefficients as functions in Q, the set of rational numbers. The following calculations show how addition and multiplication of functions f and g are defined.

Addition

$x + 1: 4 \to 5$	$f: a \to f(a)$
$x + 2: 4 \to 6$	$g: a \to g(a)$
$2x + 3: 4 \to 11$	$f + g: a \to f(a) + g(a)$

Multiplication

$x + 1: 4 \to 5$	$f: a \to f(a)$
$x + 2: 4 \to 6$	$g: a \to g(a)$
$(x + 1) \cdot (x + 2): 4 \to 30$	$f \cdot g: a \to f(a) \cdot g(a)$

Exercises 6–3

1. Let $f = 2x + 1$, $g = x + 3$. (a) Compute, as you did in high school, the functions $f + g$ and $f \cdot g$. (b) Show for your sum and product that $f + g$: $5 \to f(5) + g(5)$, and $f \cdot g$: $5 \to f(5) \cdot g(5)$.

2. Let $f = x^2 - 2x + 3$, $g = -x^2 + 2x - 3$. (a) Compute $f(4)$ and $g(4)$. (b) Show that $f + g$: $a \to 0$ for every $a \in Q$.

3. The functions f and g of Exercise 2 are said to be *negatives* of each other. One might write their sum

   $$f + g \quad \text{as} \quad 0 \cdot x^2 + 0 \cdot x + 0.$$

   This special polynomial has the property that it maps every rational number upon zero.

   $$0 \cdot x^2 + 0 \cdot x + 0: a \to 0, \qquad \forall a \in Q.$$

   We denote this polynomial by $\bar{0}$ and call it the *zero* polynomial function.

   Prove that if f is any polynomial function and x is any rational number, then

   $$(f + \bar{0}): x \to f(x), \quad \text{and so} \quad f + \bar{0} = f.$$

4. Let $f = x^2 - 2x + 3$ and $g = -x^2 + 2x - 2$. Then $f + g = 0 \cdot x^2 + 0 \cdot x + 1$. Prove that for all $a \in Q$,

   $$f + g: a \to 1.$$

5. We denote the special polynomial of Exercise 4 by $\bar{1}$. It has the property that it maps every rational number upon 1. Prove that for every polynomial function f,

   $$f \cdot \bar{1}: x \to f(x), \qquad \forall x \in Q,$$

   and hence $f \cdot \bar{1} = f$.

6. Verify that the system of polynomial functions in Q, together with the operations addition and multiplication, have the following properties:

a) Addition and multiplication of polynomial functions are both binary operations in this set of functions.

b) Addition has the commutative and associative properties.

c) Multiplication is commutative and associative.

d) Multiplication is distributive over addition.

e) The function $\bar{0}$ is the identity for addition.

f) Each function f has a unique negative $-f$ such that $f + (-f) = \bar{0}$.

g) The function $\bar{1}$ is the identity for multiplication.

Exercise 6 in the above list calls attention to the fact that the set of polynomial functions over the rationals is a *ring* relative to the operations, addition and multiplication of polynomials. Note that this ring is not a *field*. The multiplicative inverse of a polynomial function is in general not a polynomial function. Note that if $a \neq 2$,

$$\frac{1}{x - 2} \cdot (x - 2): a \rightarrow \frac{1}{a - 2} \cdot (a - 2) = 1.$$

This indicates that the product of the function

$$\frac{1}{x - 2}$$

and the polynomial function $x - 2$ is $\bar{1}$. The product of these functions maps every rational number (except 2) upon 1. We do refer to $1/(x - 2)$ as the reciprocal (multiplicative inverse) of $x - 2$, but $1/(x - 2)$ is not a polynomial function. We call $1/(x - 2)$ a *rational* function. Just as one can pass from the *ring I* of integers to the *field Q* of rational numbers by forming "quotients" of integers, one can pass from a *ring* of polynomial functions to a *field* of rational functions by forming quotients of polynomials.

The ring of polynomial functions that we have considered here is an *algebra* over Q. The phrase "algebra over Q" indicates that besides the operations of addition and multiplication, there is a third binary operation involving polynomials and rational numbers. You use this operation when you compute

$$3(2x + 1) = 6x + 3$$

"multiplying" the *polynomial function* $2x + 1$ by the *rational number* 3. Interestingly, school algebra is largely the study of two algebras, the algebra of polynomial functions and the algebra of rational functions. We have considered polynomials with rational coefficients but of course one can consider polynomials over the real number system R or over the complex number system C. Examples are:

$$\sqrt{2}\, x^2 + \pi x + \tfrac{1}{3}, \qquad (3 + i)x^2 + \sqrt{7}\, x + \tfrac{7}{5}.$$

Note that whether or not a polynomial can be factored into a product of lower degree polynomials may depend upon its coefficient field. Over Q the polynomial $x^2 - 2$ cannot be factored into the product of two first-degree polynomials. We say that $x^2 - 2$ is *prime* over Q. But over R this polynomial can be factored:

$$x^2 - 2 = (x + \sqrt{2})(x - \sqrt{2}).$$

The polynomial $x^2 + 1$ is prime over R (Can you prove this?), but over the field C of complex numbers, since $i^2 = -1$ we have

$$x^2 + 1 = (x + i)(x - i).$$

It is instructive to consider polynomial functions over finite systems. For example, mod-3 arithmetic provides us with a finite field of three elements. The tables for $+$ and $\cdot$ are given below.

+	0	1	2
0	0	1	2
1	1	2	0
2	2	0	1

$\cdot$	0	1	2
0	0	0	0
1	0	1	2
2	0	2	1

Over this 3-element field there are three constant polynomial functions, $\bar{0}$, $\bar{1}$, and $\bar{2}$.

$$0 \xrightarrow{\bar{0}} 0, \qquad 1 \xrightarrow{\bar{0}} 0, \qquad 2 \xrightarrow{\bar{0}} 0,$$
$$0 \xrightarrow{\bar{1}} 1, \qquad 1 \xrightarrow{\bar{1}} 1, \qquad 2 \xrightarrow{\bar{1}} 1,$$
$$0 \xrightarrow{\bar{2}} 2, \qquad 1 \xrightarrow{\bar{2}} 2, \qquad 2 \xrightarrow{\bar{2}} 2.$$

There are six linear polynomial functions:

$$x, \qquad x + 1, \qquad x + 2, \qquad 2x, \qquad 2x + 1, \qquad 2x + 2.$$

Check the calculations below.

$$0 \xrightarrow{x} 0, \qquad 1 \xrightarrow{x} 1, \qquad 2 \xrightarrow{x} 2,$$
$$0 \xrightarrow{x+1} 1, \qquad 1 \xrightarrow{x+1} 2, \qquad 2 \xrightarrow{x+1} 0,$$
$$0 \xrightarrow{x+2} 2, \qquad 1 \xrightarrow{x+2} 0, \qquad 2 \xrightarrow{x+2} 1,$$
$$0 \xrightarrow{2x} 0, \qquad 1 \xrightarrow{2x} 2, \qquad 2 \xrightarrow{2x} 1,$$
$$0 \xrightarrow{2x+1} 1, \qquad 1 \xrightarrow{2x+1} 0, \qquad 2 \xrightarrow{2x+1} 2,$$
$$0 \xrightarrow{2x+2} 2, \qquad 1 \xrightarrow{2x+2} 1, \qquad 2 \xrightarrow{2x+2} 0.$$

There are eighteen second-degree polynomial functions.

$$x^2, \ x^2 + 1, \ x^2 + 2, \ x^2 + x, \ x^2 + x + 1, \ldots, \ 2x^2, \ 2x^2 + 1, \ldots, \ 2x^2 + 2x + 2.$$

It is interesting to graph these functions. Our coordinate plane consists of only nine points.

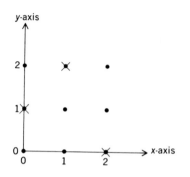

Graph of the polynomial function
$x + 1 = \{(0,1),(1,2),(2,0)\}$

The graph of each polynomial function will be a set of points $\{(0, a), (1, b), (2, c)\}$, where there are three possible choices for each of a, b, c. Hence there are only 27 possible polynomial functions over our field of three elements: 3 constant polynomials, 6 of first degree, and 18 of second degree. It is a natural conjecture that these 27 are all different from one another, and hence exhaust the set of all polynomial functions. This is indeed the case. Note that the polynomial function x^3 has the graph

$$x^3 \colon \{(0, 0), (1, 1), (2, 2)\},$$

and this is identical with the graph of the function x,

$$x \colon \{(0, 0), (1, 1), (2, 2)\}.$$

Hence we can write

$$x^3 = x$$

in the sense that the symbols x^3 and x name the same function.

Since the functions x^3 and x are the same, we can simplify a function like

$$x^4 + x^3 + 2x^2 + x + 1$$

by replacing every occurrence of x^3 by x. There is no need to consider polynomial functions of degree higher than 2.

$$x^4 + x^3 + 2x^2 + x + 1 = x \cdot x^3 + x^3 + 2x^2 + x + 1$$
$$= x \cdot x + x + 2x^2 + x + 1 = 2x + 1,$$

since $x^2 + 2x^2 = 0 \cdot x^2$.

The exercises below refer to polynomials over the field of three elements.

Exercises 6–4

1. Show that there are 54 third-degree polynomial expressions. Of course, each of these represents the same polynomial function as does some one of our earlier 27 expressions. (Can you show that these 54 fall into 27 "equal" pairs?)

2. Show that each of the polynomial functions $x^3 + 2x$ and $2x^3 + x$ is the polynomial function $\bar{0}$, and that these are the only two such third-degree polynomials. Relate this result to the question of Exercise 1.

3. Express $2x^5 + x^4 + x^3 + 2x^2 + x + 1$ as a polynomial function of degree 2 or less.

4. Show that the function $x^2 + 1$ is its own multiplicative inverse, i.e.,

$$(x^2 + 1) \cdot (x^2 + 1) = \bar{1}.$$

5. Show that $x + 1$ has no multiplicative inverse.

6. Show that $x^2 + x + 2$ and $2x^2 + 2x + 1$ are negatives of each other, i.e., their sum is $\bar{0}$. Show also that $x^3 + 2$ and $2x + 1$ are negatives of each other.

7. Observe that some second-degree polynomial expressions can be factored:
   $x^2 = x \cdot x$; $x^2 + 2 = (x + 1)(x + 2)$; $x^2 + x + 1 = (x + 2)(x + 2)$;   ....
   Others, like $x^2 + 1$, cannot be factored. In the set of nine quadratic polynomials below there are three that cannot be factored in our mod-3 system. Find them.

   a) x^2 b) $x^2 + 1$ c) $x^2 + 2$
   d) $x^2 + x$ e) $x^2 + x + 1$ f) $x^2 + x + 2$
   g) $x^2 + 2x$ h) $x^2 + 2x + 1$ i) $x^2 + 2x + 2$

8. Any second-degree equation with leading coefficient 2, as $2x^2 + x + 2 = 0$, can be transformed into an equivalent equation with leading coefficient 1 by multiplying by 2.

   $$2x^2 + x + 2 = 0 \to 2(2x^2 + x + 2) = 0 \to x^2 + 2x + 1 = 0.$$

   In this case, since the polynomial $x^2 + 2x + 1$ factors into $(x + 1)(x + 1)$, the equation has roots in our system. Give the solution set for this equation.

9. Show that each of the three prime polynomials that you located in Exercise 7 has the following properties.

   a) The polynomial has no root in the mod-3 system.

   b) The polynomial is its own multiplicative inverse.

Whenever we consider an algebra over some number system, one of our basic problems is the solution of polynomial equations. If we encounter an equation that has no solution in our system, then it is natural to ask ourselves whether we can invent a larger system in which this equation has a solution. In the set W, equations of the form

$$x + a = 0, \qquad a \in W$$

have no solution unless $a = 0$. We remove this "defect" by inventing solutions denoted by ^{-}a. These new numbers, combined with the old numbers of W, furnish us with our set I.

However, I linear equations of the form

$$ax - b = 0$$

have no solution unless the integer a is a factor of b. For each such equation we agree to denote a root by the symbol b/a and agree that

$$a \cdot \frac{b}{a} = b.$$

The adjunction of these numbers to I leads to the rational number system Q. In Q every linear equation

$$rx + s = 0$$

with

$$r, s \in Q \qquad \text{and} \qquad r \neq 0,$$

has a solution in Q.

But, when we look at quadratic equations over Q we find very simple ones which have no root in Q. The classic examples are

$$x^2 - 2 = 0, \qquad x^2 - 3 = 0, \qquad \dots .$$

Here the ideas begin to get difficult. We sense that it would be desirable to invent enough new numbers so that every polynomial equation of degree one or more would have roots in our new system. This goal leads us finally through the real number system R to the complex number system C. And here our objective is attained, for every polynomial equation of degree one or more with complex number coefficients has a complex root. One of the achievements of classical algebra is the description of this complex number system and the proof that it has this property of *algebraic completeness*. We cannot concern ourselves here with these details, but in the next section we shall study a much simpler related topic. We shall show how to extend our finite field of three elements by adjoining to it new numbers which are roots of the three prime polynomials which you located in Exercise 7 of the above exercise list.

6–3 Algebraic Extensions

In this section we show how a field can be enlarged by adjoining a root of a polynomial equation. We shall illustrate the technique using the finite field of three elements. A familiar analog of our procedure is the adjunction of $\sqrt{2}$ to the rational number system. We invented the number $\sqrt{2}$ in order to have a root of the equation $x^2 = 2$ which has no rational root. Then using the commutative, associative, etc., laws, we computed with this new number as below:

$$(2 + 3\sqrt{2}) + (1 + 2\sqrt{2}) = 3 + 5\sqrt{2}, \qquad \sqrt{2}\,(3 + \sqrt{2}) = 2 + 3\sqrt{2},$$

$$\frac{1 + \sqrt{2}}{2 - \sqrt{2}} = \frac{(1 + \sqrt{2})(2 + \sqrt{2})}{(2 - \sqrt{2})(2 + \sqrt{2})} = \frac{4 + 3\sqrt{2}}{2} = 2 + \frac{3}{2}\sqrt{2}.$$

In Exercise 7 of Exercises 6–4, you determined that the following equations have no roots in our field of three elements.

$$x^2 + 1 = 0,$$

$$x^2 + x + 2 = 0,$$

$$x^2 + 2x + 2 = 0.$$

The first of these equations can be written as

$$x^2 = 2,$$

and so, just as in Q, there is no number in our field whose square is 2:

$$0^2 \neq 2, \qquad 1^2 \neq 2, \qquad 2^2 \neq 2.$$

We could introduce a root of this equation, but instead we shall use the equation

$$x^2 + x + 2 = 0.$$

We invent a new number and agree that it satisfies this equation. We denote this number by t. Hence

$$t^2 + t + 2 = 0.$$

We adjoin t to our field of three elements, agreeing, of course, that

$$1 \cdot t = t, \qquad 0 \cdot t = 0, \qquad 2(1 + t) = 2 + 2t, \qquad \ldots,$$

i.e., we agree that our old familiar properties hold for our new set of numbers

$$\{0, 1, 2, t, 1 + t, 2 + t, 2t, \ldots\}.$$

As examples of computation in this new set:

$$t + t = 1 \cdot t + 1 \cdot t = (1 + 1) \cdot t = 2t,$$
$$2t + t = (2 + 1) \cdot t = 0 \cdot t = 0.$$

Note that since $t^2 + t + 2 = 0$, we have

$$t^2 + t + 2 + (2t + 1) = 2t + 1,$$
$$t^2 + (2t + t) + (2 + 1) = 2t + 1,$$
$$t^2 = 2t + 1.$$

Hence in any expression involving t^2 we may replace t^2 by $2t + 1$.

Let us list elements in the new system that we obtain by adjoining t to our field and repeatedly adding and multiplying any numbers.

$$S = \{0, 1, 2, t, t + 1, t + 2, 2t, 2t + 1, 2t + 2\}.$$

Note that only the nine numbers listed above are obtained. This new set S of nine numbers is closed under addition and multiplication. For example,

$$(t + 1)(2t + 1) = 2t^2 + t + 2t + 1 = 2t^2 + 1 = 2(2t + 1) + 1 = t.$$

Our task now is to study the structure of this new system S. Is it a group under addition? Do the nonzero elements form a group under multiplication? If the answer to both these questions is Yes, then S is a field. The next set of exercises provides an opportunity to study S in detail.

Exercises 6–5

1. We should verify that the nine elements in

   $$S = \{0, 1, 2, t, t + 1, t + 2, 2t, 2t + 1, 2t + 2\}$$

   are pairwise distinct. Of course, we know that $t \neq 0$, $t \neq 1$, and $t \neq 2$, for t is a root of the equation $x^2 + x + 2 = 0$, and no one of 0, 1, 2 satisfies this equation. Prove that $t \neq 2t + 1$.

2. Show that $t + 1$ and $2t + 2$ are negatives of each other. What is the negative of $t + 2$? Has every number in S a negative? Is S a group relative to addition?

3. Construct the addition table for S and point out how this table shows that each number has a negative in S.

4. Verify that $t(t + 1) = 1$. This shows that t and $t + 1$ are reciprocals (multiplicative inverses) of each other. Show that $2t$ and $2t + 2$ are reciprocals as are $2t + 1$ and $t + 2$. Do the nonzero elements of S form a group relative to multiplication? Is S a field?

5. For reference, the multiplication table for the nonzero elements of S is given at the top of the following page. Verify all entries in the fourth and fifth rows of this table.

·	1	2	t	$t+1$	$t+2$	$2t$	$2t+1$	$2t+2$
1	1	2	t	$t+1$	$t+2$	$2t$	$2t+1$	$2t+2$
2	2	1	$2t$	$2t+2$	$2t+1$	t	$t+2$	$t+1$
t	t	$2t$	$2t+1$	1	$t+1$	$t+2$	$2t+2$	2
$t+1$	$t+1$	$2t+2$	1	$t+2$	$2t$	2	t	$2t+1$
$t+2$	$t+2$	$2t+1$	$t+1$	$2t$	2	$2t+2$	1	t
$2t$	$2t$	t	$t+2$	2	$2t+2$	$2t+1$	$t+1$	1
$2t+1$	$2t+1$	$t+2$	$2t+2$	t	1	$t+1$	2	$2t$
$2t+2$	$2t+2$	$t+1$	2	$2t+1$	t	1	$2t$	$t+2$

6. Explain why in the above table the $(2t+2)$-row is the double of the $(t+1)$-row, and the $(t+1)$-row is the double of the $(2t+2)$-row.

7. Using the table, compute:

a) $(2t+1)(2t+2)(t+2)$ b) $t(t+1)(t+2)$

c) $(t+1)(t+2)(2t+1)$ d) $2t(t+1)(2t+2)$

e) t^2 f) t^3 g) t^4 h) t^5 i) t^6 j) t^7 k) t^8

l) $(t+2)^2$ m) $(t+2)^4$ n) $t \div (t+1)$

o) $(t+1) \div t$ p) $1 \div (t+2)$ q) $(2t+1) \div (2t+2)$

8. Recall that we invented t in order to have a root for the equation

$$x^2 + x + 2 = 0.$$

Show that $2t + 2$ is also a root of this equation.

9. Besides the equation

$$x^2 + x + 2 = 0,$$

the equations

$$x^2 = 2 \quad \text{and} \quad x^2 + 2x + 2 = 0$$

had no roots in our three-element field. Show that each of these equations has two roots in S, and so all quadratic equations with coefficients in the three-element field have roots in S.

10. Show that over S, the three polynomials

   $$x^2 + 1, \qquad x^2 + x + 2, \qquad \text{and} \qquad x^2 + 2x + 2,$$

   which could not be factored over our three-element field, factor as:

   a) $x^2 + x + 2 = (x + 2t)(x + t + 1)$,

   b) $x^2 + 1 = (x + 2t + 1)(x + t + 2)$,

   c) $x^2 + 2x + 2 = (x + t)(x + 2t + 2)$.

11. Parts (e) through (k) of Exercise 7 show that every nonzero number in S is a power of t.

   $$2t + 1 = t^2, \qquad 2t + 2 = t^3, \qquad 2 = t^4, \qquad 2t = t^5,$$
   $$t + 2 = t^6, \qquad t + 1 = t^7, \qquad 1 = t^8.$$

   Use this fact to show that every one of the nine numbers in S is a root of the equation $x^9 + 2x = 0$.

12. Use the table of powers of t in Exercise 11 to solve the following linear equations:

   a) $(t + 2)x = 2t + 2$,   b) $tx = 1$,

   c) $(t + 1)x = t + 2$,   d) $(2t + 1)x = t + 2$,

   e) $tx + t + 1 = 2t$,   f) $(2t + 2)x + t + 2 = 0$.

13. Since S is a field, fractional expressions like

   $$\frac{1}{t + 2} \qquad \text{and} \qquad \frac{2t + 1}{t}$$

   represent elements of S. The number $1/(t + 2)$ is the reciprocal of $t + 2$, i.e.,

   $$\frac{1}{t + 2} = 2t + 1$$

   and

   $$\frac{2t + 1}{t} = (2t + 1) \cdot \frac{1}{t} = (2t + 1)(t + 1), \text{ etc.}$$

   Perform the computations below by any method you like.

   a) $\dfrac{t + 2}{t + 1} + \dfrac{2}{t + 1}$ b) $\dfrac{2t + 2}{t + 2} \cdot \dfrac{t + 2}{t + 1}$

   c) $t + \dfrac{2t + 1}{t + 2}$ d) $\dfrac{1}{t} + \dfrac{t}{t + 1}$

14. We may consider polynomial functions *over* S. For example, there are nine constant polynomial functions. We list some of the linear and quadratic

polynomials below:

x, $x + 1$, $x + 2$, $x + t$, $x + t + 1$, ..., $(2t + 2)x + 2t$, $(2t + 2)x + 2t + 1$, $(2t + 2)x + 2t + 2$, x^2, $x^2 + 1$, ..., $x^2 + tx + 2$, $x^2 + tx + t$, $x^2 + tx + t + 1$, ..., $tx^2 + (2t + 1)x + 2$, ..., $(2t + 2)x^2 + (2t + 2)x + 2t + 2$.

a) How many first-degree polynomials are there over S?

b) How many second-degree polynomials over S?

c) Show that the polynomial function x^9 is the identity function that maps each element of S upon itself, that is,

$\forall x \in S$, $x^9 = x$.

d) If polynomial functions over S are graphed, how many "points" are there in the coordinate plane?

e) How many different polynomial functions can there be in S?

As the reader may have anticipated, it is easy to find polynomial equations of second degree (and higher) which have coefficients in S but have no root in S. For example, the equation

$x^2 = t$ or $x^2 + 2t = 0$

has no root in S. This can be seen to be true by looking down a diagonal of the multiplication table for S. (Explain.) The only numbers in S which are squares are

1, 2, $2t + 1$, and $t + 2$.

(Use the facts that every nonzero number in S is some power of t, and that $t^8 = 1$ to explain this.) If we wish, then, we can invent a root of this equation and tack it on to S. If we denote this root by $\sqrt{t}$, then we would obtain a new system of 81 numbers.

$0, 1, 2, t, \ldots, t + 2, \sqrt{t}, \sqrt{t} + 1, \sqrt{t} + 2, \sqrt{t} + t, \ldots, (2t + 2)\sqrt{t} + (2t + 2)$.

It is interesting that the invention of this single new number $\sqrt{t}$ enables us to solve *every* quadratic equation with coefficients in S. There are exactly 36 equations like

$x^2 + 2t = 0$

which have no roots in S, and the invention of $\sqrt{t}$ provides us with 72 new numbers, just enough to provide 2 roots for each of these equations. There was no need to use the special equation $x^2 + 2t = 0$. We could choose any one of the 36 quadratic equations without roots in S, and the invention of a root for this equation would enable us to solve every quadratic equation with coefficients in S.

The process of building on and on to these finite fields by locating equations without roots, inventing a root, and adjoining the root to form a larger field can be continued indefinitely. Thus we have a technique for constructing infinitely many finite fields. We have touched upon only a few of the mathematical ideas associated with these fields. The interested reader will be able to formulate many problems of his own. We indicate below how any quadratic equation over S can be solved once we possess the new number $\sqrt{t}$. We consider the special equation

$$x^2 + 2tx + t = 0,$$

but the reader will note that the method applies to any quadratic.

1. $x^2 + 2tx + t = 0$.

Now note that $(x + t)^2 = x^2 + 2tx + 2t + 1$. Hence we can rewrite (1) as

2. $(x + t)^2 = t + 1$. (Explain.)

Also, $t + 1 = t^7$. (See Exercise 11 above.)

3. $(x + t)^2 = t^7$.

4. $x + t = t^3 \cdot \sqrt{t}$ or $2t^3 \cdot \sqrt{t}$. (Why?)

Now $t^3 = 2t + 2$,   so

5. $x + t = (2t + 2)\sqrt{t}$ or $x + t = (t + 1)\sqrt{t}$.

6. $x = (2t + 2)\sqrt{t} + 2t$ or $x = (t + 1)\sqrt{t} + 2t$.

Solve similarly the quadratic $x^2 + (2t + 2)x + 1 = 0$.

7 Similarity

7-1 Introduction

Intuitively we know that two geometric figures are *similar* if they have the same shape. A photograph that does not distort presents an image similar to the object photographed. Any two segments are similar, every circle is similar to every other circle, and every square to every other square. Is every rectangle similar to every other rectangle? Explain.

Just as with congruence we can associate similarity with concepts of motion. For example, we can think of an inflated balloon as a perfect sphere, but if we introduce more air into the balloon, it "moves" (expands) into the position of a larger sphere similar to the original one. Ideas of similarity are exemplified by blueprints and maps. Two maps of the same state are similar. Give other examples of similar figures.

We need to make the concept "same shape" more precise. In each pair shown below, the two figures do not have the same shape. Explain why.

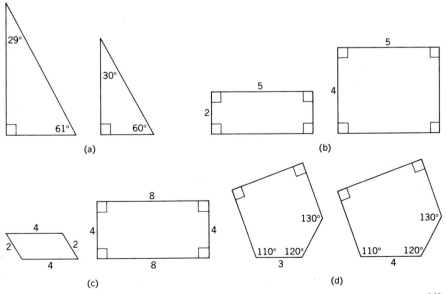

(a) (b) (c) (d)

The previous examples suggest that a definition of *similarity of two figures* might involve both *distances* and *angles*. Explain why you think the pairs of figures shown below are similar.

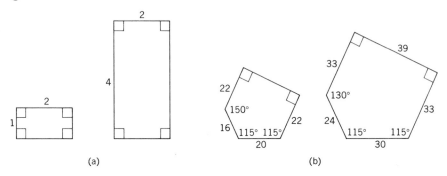

(a)   (b)

Exercises 7–1

1. If two polygons are similar, must it be possible to match the angles of one polygon with the angles of the other so that corresponding angles are congruent?

2. If, for two polygons, it is possible to match the angles of one with the angles of the other so that corresponding angles are congruent, is it necessarily the case that the polygons are similar? Support your answer with an example.

3. If the sides of one polygon can be matched with the sides of a second so that each side of the first is three times as long as the corresponding side of the second, are the polygons necessarily similar? Support your answer with an example.

4. Are two equilateral triangles necessarily similar? Explain your answer.

5. In the figure, M and N are midpoints of $\overline{AB}$ and $\overline{AC}$. Explain why $\triangle AMN$ is similar to $\triangle ABC$.

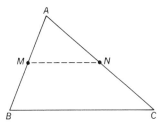

7–2 Ratio and Proportion

The problems of Section 7–1 call attention to the fact that if a, b, c, ... and a', b', c', ... are the lengths of corresponding sides of two similar polygons then not only must corresponding angles be congruent but the ratios

$$\frac{a}{a'}, \frac{b}{b'}, \frac{c}{c'}, \ldots$$

must all be equal. This relationship is abbreviated as

$$a, b, c, \ldots : a', b', c', \ldots$$

and we say the ordered set of numbers $(a, b, c, \ldots)$ is *proportional* to the ordered set $(a', b', c', \ldots)$. An equation of the form

$$\frac{a}{b} = \frac{c}{d}$$

is called a *proportion*.

To say that four numbers a, b, c, d are *proportional* is to assert that the equation above holds.

A basic and useful property of proportions is the following:

$$\frac{a}{b} = \frac{c}{d} \qquad \textit{if and only if} \qquad ad = bc.$$

Give a proof of this relationship.

Exercises 7–2

1. Write two sequences, each of five positive numbers, that are proportional.

2. Show that

$$\text{if } \frac{a}{b} = \frac{c}{d}, \qquad \text{then} \quad \frac{b}{a} = \frac{d}{c}.$$

3. Show that

$$\text{if } \frac{a}{b} = \frac{c}{d}, \qquad \text{then} \quad \frac{a}{c} = \frac{b}{d}.$$

4. Show that

$$\text{if } \frac{a}{b} = \frac{c}{d}, \qquad \text{then} \quad \frac{c}{a} = \frac{d}{b}.$$

5. Choose a, b, c so that the sets $(4, a, 10, b)$ and $(6, 12, c, 15)$ are proportional.

6. Use the fact that

$$\text{if } \frac{a}{b} = \frac{c}{d}, \qquad \text{then} \quad \frac{a}{b} + 1 = \frac{c}{d} + 1,$$

to establish the following relationships:

a) If $\dfrac{a}{b} = \dfrac{c}{d}$,   then $\dfrac{a+b}{b} = \dfrac{c+d}{d}$,

b) If $\dfrac{a}{b} = \dfrac{c}{d}$,   then $\dfrac{b}{a+b} = \dfrac{d}{c+d}$.

7. Complete the following statements.

a) If $\dfrac{a}{b} = \dfrac{c}{d} = \dfrac{e}{f}$,   then $\dfrac{a+b}{a} =$ _____ $=$ _____.

b) If $\dfrac{a}{b} = \dfrac{c}{d}$,   then $\dfrac{a-b}{b} =$ _____.

c) If $\dfrac{a}{b} = \dfrac{c}{d}$,   then $\dfrac{a-b}{a+b} =$ _____.

8. Prove that

if $\dfrac{a}{b} = \dfrac{c}{d}$,   then $\dfrac{a+c}{b+d} = \dfrac{a}{b}$.

9. Consider the continued proportion

$$\tfrac{2}{3} = \tfrac{4}{6} = \tfrac{6}{9} = \tfrac{8}{12} = \tfrac{18}{27}.$$

Verify that

$$\frac{2 + 4 + 6 + 8 + 18}{3 + 6 + 9 + 12 + 27} = \frac{2}{3}.$$

Generalize the results of the above for proportional sequences $a, b, c, d, \ldots$: $a', b', c', d', \ldots$.

10. Complete each statement.

a) If $3a = 2b$,   then $\dfrac{a}{b} =$ _____,   and $\dfrac{a}{2} =$ _____.

b) If $4m = 15$,   then $\dfrac{m}{5} =$ _____,   and $\dfrac{m}{3} =$ _____.

c) If $6x = 5 \cdot 9$,   then $\dfrac{x}{5} =$ _____,   and $\dfrac{5}{x} =$ _____.

d) If $\dfrac{2a}{3b} = \dfrac{7c}{5d}$,   then $\dfrac{a}{b} =$ _____   and $\dfrac{b}{a} =$ _____.

11. Consider the three sequences below. Which pairs of sequences are proportional?

a) 3, 8, 12, 17   b) 9, 24, 36, 51   c) $\tfrac{7}{2}, \tfrac{28}{3}, 14, \tfrac{119}{6}$

12. If a, b, c are positive numbers and

$$\frac{a}{b} = \frac{b}{c},$$

then b is called the *geometric mean* of a and c. The *arithmetic mean* of a and c is

$$\frac{a + c}{2}.$$

a) Obtain a formula for the geometric mean of two positive numbers.

b) Are the arithmetic and geometric means of two distinct positive numbers always between the numbers?

13. Find both the arithmetic and geometric means for each pair of numbers.

a) 2 and 8   b) 3 and 12   c) 5 and 45

d) 4 and 9   e) 9 and 16   f) 12 and 15

14. If $a \neq b$, how do the arithmetic and geometric means of a and b compare?

15. Triangles ABC and DEF are similar with a, b, c: d, e, f and

$$\angle A \cong \angle D, \qquad \angle B \cong \angle E, \qquad \angle C \cong \angle F.$$

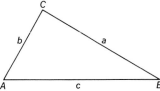 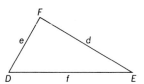

Given that:

a) $b = 3$, $d = 4$, $e = 12$. Find a.

b) $a = 6$, $d = 15$, $c = 2$. Find f.

c) $b = 5$, $c = 2$, $f = 3$. Find e.

d) $d = 4$, $b = 2$, $e = 7$. Find a.

e) $f = \frac{9}{2}$, $d = 5$, $a = 6$. Find c.

f) $b = e$, $d = 1.32$. Find a.

g) $d = 3$, $a = 2$, $e = 1$. Find b.

h) $c = 2$, $a = 2.1$, $f = 3.4$. Find d.

i) $d = \frac{5}{2}$, $c = \frac{10}{3}$, $f = \frac{17}{4}$. Find a.

7–3 Similar Triangles

The statement that two triangles are similar is abbreviated as

$$\triangle ABC \sim \triangle DEF.$$

This conveys the information that

$$\angle A \cong \angle D, \qquad \angle B \cong \angle E, \qquad \angle C \cong \angle F,$$

and

$$\frac{m(\overline{AB})}{m(\overline{DE})} = \frac{m(\overline{BC})}{m(\overline{EF})} = \frac{m(\overline{CA})}{m(\overline{FD})}.$$

We say,

Corresponding angles are *congruent* and corresponding sides are *proportional*.

In this section we will investigate minimal conditions for two triangles to be similar. The question as to whether two polygons are similar can be reduced to one of similarity of triangles.

Two polygons are similar if and only if they can be partitioned into the same number of triangles which are similar each to each and "similarly placed."

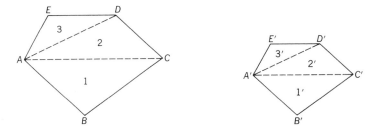

Using a protractor construct two triangles, each having angles of 40°, 20°, and 120°. For the first triangle, make the side opposite the 120° angle 6 in. long; for the second, make the side opposite the 120° angle 2 in. long. Measure and compare the other pairs of sides.

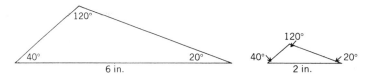

Your measurements should suggest the theorem that if the angles of two triangles are congruent, then the sides are proportional and the triangles are similar. There is no simple proof for this theorem. Any proof rests essentially upon sophisticated properties of the real number system. In the next list of exercises we indicate how two proofs of this theorem can be made. One depends upon properties of parallel lines; the other, upon area concepts.

Exercises 7–3

1. In Problem 5 of Exercises 7–1, you observed that the line joining midpoints of the sides of a triangle cuts off a small triangle similar to the large one.

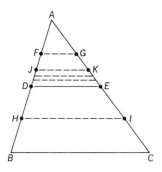

$$\frac{m(\overline{AD})}{m(\overline{AB})} = \frac{m(\overline{AE})}{m(\overline{AC})} = \frac{m(\overline{DE})}{m(\overline{BC})} = \frac{1}{2}.$$

a) Show that if F and G are midpoints of $\overline{AD}$ and $\overline{AE}$, then

$$\triangle AFG \sim \triangle ABC \qquad \text{and} \qquad \frac{m(\overline{FG})}{m(\overline{BC})} = \frac{1}{4}.$$

b) Show that if H and I are midpoints of $\overline{DB}$ and $\overline{EC}$, then

$$\triangle AHI \sim \triangle ABC.$$

c) Show that if J and K are midpoints of $\overline{FD}$ and $\overline{GE}$, then

$$\triangle AJK \sim \triangle ABC \qquad \text{and} \qquad \frac{m(\overline{JK})}{m(\overline{BC})} = \frac{3}{8}.$$

2. As shown in the figure, let $\overline{XY}$ be any segment parallel to $\overline{BC}$. Explain how we can use the method of Exercise 1 to construct parallels to $\overline{BC}$ as "close as we wish" to $\overline{XY}$, cutting off triangles similar to $\triangle ABC$. Explain how "common sense" suggests that $\triangle AXY \sim \triangle ABC$. Limit notions are required to make a rigorous proof.

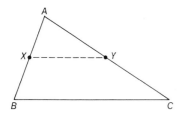

3. Now assume that any line parallel to the base of a triangle and cutting the other sides cuts off a triangle similar to the original triangle. With this assumption we can prove that any two triangles with congruent angles are similar.

   In triangles ABC and DEF, let $\angle A \cong \angle D$, $\angle B \cong \angle E$, and $\angle C \cong \angle F$. Let $\overline{AB} > \overline{DE}$. Choose E' as shown so that $\overline{AE'} \cong \overline{DE}$, and draw $\overline{E'F'}$ parallel to $\overline{BC}$.

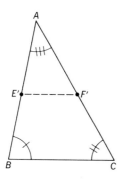 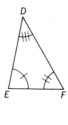

   a) Why is $\triangle AE'F' \cong \triangle DEF$?

   b) Why is $\triangle AE'F' \sim \triangle ABC$?

   c) Why is $\triangle DEF \sim \triangle ABC$?

4. Assuming the results of Exercise 3, argue that two triangles are similar if two angles of one are respectively congruent to two angles of the other.

5. In this exercise we develop an *area* proof of the basic theorem that a line parallel to the base of a triangle cuts off a triangle similar to the original. In $\triangle ABC$, $\overline{DE} \parallel \overline{BC}$. We shall show that

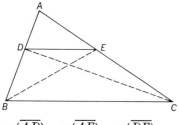

$$\frac{m(\overline{AD})}{m(\overline{DB})} = \frac{m(\overline{AE})}{m(\overline{AC})} = \frac{m(\overline{DE})}{m(\overline{BC})}.$$

   Of course the angles of the two triangles are congruent. Why?

   a) $\triangle$'s BDE and CDE have equal areas. (Why?)

   b) $\triangle$'s BAE and CDA have equal areas. (Why?)

   c) The ratio of the area of $\triangle BAE$ to the area of $\triangle BAC$ is $\overline{AE}$ to $\overline{AC}$. (Why?)

We write this as

$$\frac{\triangle BAE}{\triangle BAC} = \frac{m(\overline{AE})}{m(\overline{AC})}.$$

d) $\dfrac{\triangle CDA}{\triangle CBA} = \dfrac{m(\overline{AD})}{m(\overline{AB})}.$ (Why?)

e) Hence

$$\frac{m(\overline{AD})}{m(\overline{AB})} = \frac{m(\overline{AE})}{m(\overline{AC})}.$$ (Why?)

f) $\dfrac{\triangle BAE}{\triangle BAC} = \dfrac{\triangle DAE}{\triangle DAC} = \dfrac{m(\overline{AE})}{m(\overline{AC})}.$ (Why?)

g) $\dfrac{\triangle BAE - \triangle DAE}{\triangle BAC - \triangle DAC} = \dfrac{m(\overline{AE})}{m(\overline{AC})} = \dfrac{\triangle BDE}{\triangle DBC} = \dfrac{m(\overline{DE})}{m(\overline{BC})}.$

h) Hence

$$\frac{m(\overline{AD})}{m(\overline{DB})} = \frac{m(\overline{AB})}{m(\overline{AC})} = \frac{m(\overline{DE})}{m(\overline{BC})}.$$

6. In the figure, $l \parallel m \parallel n \parallel p$. Name three pairs of similar triangles.

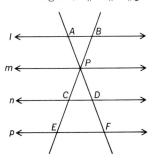

7. Prove that if $\triangle ABC \sim \triangle DEF$ and $\triangle DEF \sim \triangle GHK$, then $\triangle ABC \sim \triangle GHK$. This is the transitivity property for similarity of triangles.

8. In the figure $\overleftrightarrow{DE} \parallel \overleftrightarrow{AB}$ and $\overleftrightarrow{EF} \parallel \overleftrightarrow{AC}$. Prove that $\triangle CDE \sim \triangle EFB$.

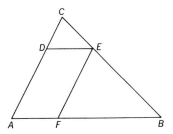

9. Planes α, β, γ are parallel. Name two pairs of similar triangles in the figure.

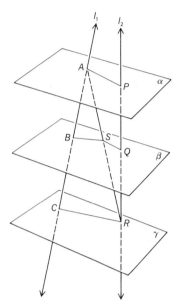

10. In the figure, lines l' and l'' are cut by parallels. The lengths of the segments are as indicated. Show that the following equations hold.

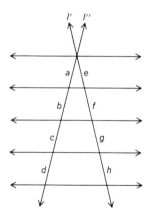

a) $\dfrac{a}{e} = \dfrac{a+b}{e+f}$,

b) $\dfrac{a}{e} = \dfrac{b}{f}$,

c) $\dfrac{a}{e} = \dfrac{b}{f} = \dfrac{c}{g} = \dfrac{d}{h} = \dfrac{b+c}{f+g} = \dfrac{c+d}{g+h}$.

We have seen that if the angles of one triangle are congruent to the angles of a second, then the sides of the triangles are proportional and the triangles are similar. (A corresponding statement for quadrilaterals is false. Why?) It is natural to ask whether proportionality of the sides of two triangles implies congruence of the angles and hence similarity. It is intuitively clear that this is so, and the proof is not difficult. We sketch the argument below.

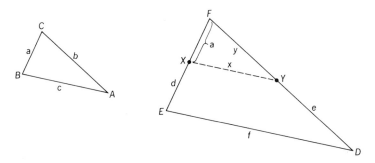

i) Let
$$\frac{a}{d} = \frac{b}{e} = \frac{c}{f} \qquad \text{with} \qquad a < d.$$

ii) Draw $\overline{XY}$ as shown, parallel to $\overline{ED}$.

iii) Then

$$\triangle FXY \sim \triangle FED \qquad \text{and} \qquad \frac{a}{d} = \frac{x}{f} = \frac{y}{e}. \quad \text{(Why?)}$$

iv) By (i) and (iii), $x = c$ and $y = b$, so $\triangle CBA \cong \triangle FXY$, and $\triangle CBA \sim \triangle FED$. (Why?)

We now have two sets of "minimal" conditions for similarity of triangles:

1) Congruence of two pairs of angles,
2) Proportionality of the sides.

We can get a third "mixed" condition.

3) Congruence of one pair of angles and proportionality of the including sides. For example, in the figure above: If we know that

$$\angle C \cong \angle F \qquad \text{and} \qquad \frac{a}{d} = \frac{b}{e},$$

then it follows that

$$\frac{c}{f} = \frac{a}{d},$$

and the triangles are similar. (Try to make this proof.) The problems below utilize these conditions for similarity.

Exercises 7–4

1. Decide whether the triangles in each pair shown below are similar. Explain your decision.

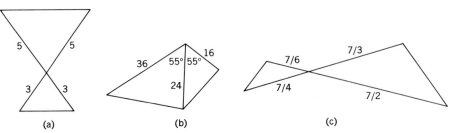

(a)   (b)   (c)

2. In the figure, x, y, and z are the lengths of MB, $\overline{MA}$, and $\overline{MC}$. There are two possible lengths of $\overline{MD}$ for which the triangles are similar. Explain, and compute these lengths.

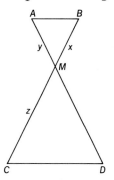

3. Lines l and m are parallel.

a) Compute the length of $\overline{AE}$, given that the lengths of $\overline{AB}$, $\overline{AD}$, and $\overline{AC}$ are respectively 9, 21, and 12.

b) Compute the length of $\overline{CE}$, given that the lengths of $\overline{AB}$, $\overline{BD}$, and $\overline{AC}$ are 9, 20, and 15, respectively.

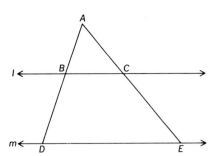

4. Three lines, concurrent at P, intersect parallel planes determining two triangles, one in each plane, as shown in the figure. The lengths of certain segments are given. Is $\triangle MHK \sim \triangle STR$? Justify your answer. Compute the lengths of $\overline{PS}$, $\overline{PT}$, $\overline{RT}$, $\overline{TS}$, and $\overline{RS}$.

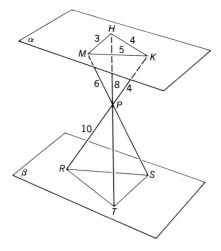

5. Decide whether the triangles in each pair are similar.

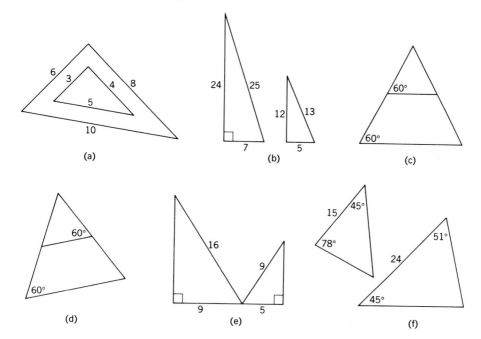

6. Is it possible for two triangles to be similar if

a) Two angles of one have measures of 60° and 70°, and two angles of the other have measures of 50° and 80°?

b) Two angles of one have measures of 45° and 75°, and two angles of the other have measures of 60° and 45°?

c) One has an angle of measure 40° and two sides of length 5, and the other has an angle of 70° and two sides of length 8?

d) One has sides of length 5, 6, and 9, and the other has a perimeter of 8420?

7. In the figure two similar quadrilaterals are partitioned into two similar triangles which are similarly placed. Describe a second way to partition these quadrilaterals into similar triangles similarly placed.

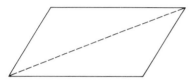 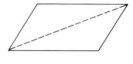

8. In the figure two quadrilaterals are partitioned into similar triangles. Explain why the quadrilaterals are not similar.

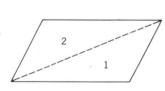 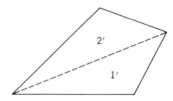

$$1 \sim 1'; \qquad 2 \sim 2'$$

9. Three segments are shown with lengths a, b, and c. Construct a fourth segment with length d such that

$$\frac{a}{b} = \frac{c}{d}.$$

[*Hint:* Construct similar triangles.]

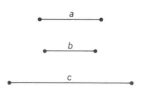

10. Divide the segment of length a into two segments whose lengths are in the ratio c to b.

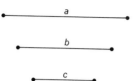

11. The figure shows that if two triangles are similar and the ratio of corresponding sides is 1 to 2, then the ratio of the *areas* of the two triangles is 1 to 4. Explain how the figure shows this.

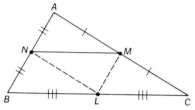

12. The figure pictures two similar triangles: $\triangle ABC \sim \triangle ADE$.
   a) What is the ratio of $\overline{AD}$ to $\overline{AB}$?
   b) What is the ratio of the area of $\triangle ADE$ to the area of $\triangle ABC$?

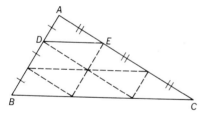

13. The figure shows two similar triangles: $\triangle ABC \sim \triangle ADE$. The ratio of $\overline{AD}$ to $\overline{AB}$ is 1 to 4. What is the ratio of the area of $\triangle ADE$ to the area of $\triangle ABC$? Sketch in lines as in Exercises 11 and 12, and verify your guess.

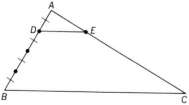

14. What is the ratio of the areas of two similar triangles if the ratio of corresponding sides is

   a) 1 to 10?   b) 1 to n?   c) 2 to 5?   d) x to y?

15. The longest sides of two similar triangles are in the ratio of 3 to 4. What is the ratio of their areas?

16. The areas of two similar triangles are in the ratio 16 to 25. What is the ratio of two corresponding sides?

17. In $\triangle ABC$, $\overline{CD}$ bisects $\angle C$. Line l is parallel to $\overline{CD}$, and side $\overline{BC}$ extended cuts l at E. Why is

   a) $\triangle BCD \sim \triangle BEA$?   b) $\dfrac{\overline{BC}}{\overline{CE}} = \dfrac{\overline{BD}}{\overline{DA}}$?   c) $\overline{CE} \cong \overline{CA}$?

   d) $\dfrac{\overline{BC}}{\overline{AC}} = \dfrac{\overline{BD}}{\overline{DA}}$?

   e) Write a theorem stating what has been proved in parts *a* through *d*.

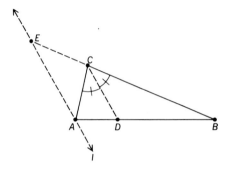

7–4 Applications of Similarity

The theory of similarity is the basis for most indirect measurement. The figures below illustrate this. A tree throws a 60-ft shadow at the same time as a 10-ft pole throws a 12-ft shadow. How tall is the tree?

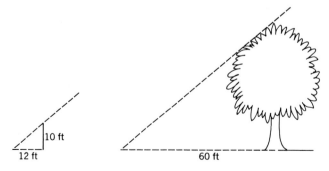

Flying at a height of 10,000 ft, a plane photographs a rectangular wheat field. It is known that at this height the photograph reduces dimensions in the ratio of 1 to 5000. Measure the dimensions of the photo and compute the approximate number of acres in the field. [There are 43,560 sq ft in 1 acre.]

A boy holds a foot rule vertically at arm's length and moves away from a building until the lines of sight through the top and bottom of the ruler run respectively to the top and base of the building. Using the distances shown in the diagram, estimate the height of the building.

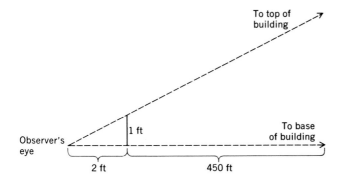

Exercises 7–5

1. Use the technique described in the last example to estimate the heights of some buildings or trees. Use either a foot rule or yardstick.

2. A man 6 ft tall casts a shadow $3\frac{1}{2}$ ft long when a building has a 16-ft shadow. How high is the building?

3. On a certain map, $\frac{3}{8}$ in. represents 50 mi. The distance between two cities is 5 in. on the map. What is the approximate airline distance between the cities?

4. To measure the distance between two points A and B on opposite sides of a river, a boy stood at A and measured the angle between the lines of sight from A to B and from A to C (another point on his side of the river). He paced off the distance from A to C, and estimated it as 100 ft. From C he measured

the angle between $\vec{CA}$ and $\vec{CB}$. Then he drew a small triangle as shown, with base 10 cm, and angles A' and C' congruent to $\angle A$ and $\angle C$. He measured $\overline{A'B'}$ and found the distance on his drawing to be about 16.5 cm. About how far is it from A to B?

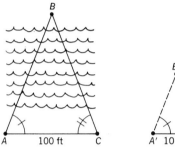

5. Explain the following proof of the Pythagorean theorem.

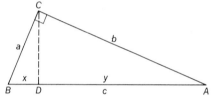

i) $\triangle DBC \sim \triangle CBA$ and $\triangle ADC \sim \triangle ACB$.

ii) $\dfrac{x}{a} = \dfrac{a}{c}$ and $\dfrac{y}{b} = \dfrac{b}{c}$.

iii) $c \cdot x = a^2$ and $c \cdot y = b^2$.

iv) $a^2 + b^2 = cx + cy = c^2$.

7–5 Trigonometry

Any two right triangles are similar if an acute angle of one is congruent to an acute angle of the other. (Why?)

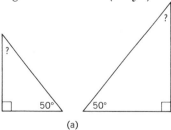
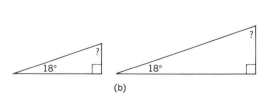

(a)   (b)

In trigonometry this fact is systematically exploited and used to solve a vast variety of problems. For convenience in solving problems, special tables of *trigonometric functions* are constructed. The three tables that we shall be concerned with are those for the

  sine, cosine, and *tangent* functions.

The three diagrams below suggest how the tables for these functions are constructed and used.

The Sine Function

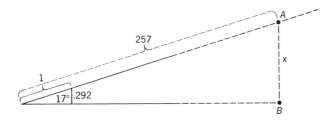

Careful measurement shows that in a 17° right triangle with hypotenuse 1, the leg *opposite* the 17° angle has a length of approximately .292. We call .292 *the sine of* 17° (abbreviated, sin 17°). Hence, in the figure the length of AB is approximately 82.5. (Explain.)

The Cosine Function

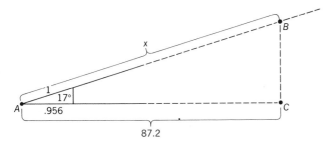

In a 17° right triangle the leg *adjacent* to the 17° angle has a length about .956 of the length of the hypotenuse. This number .956 is called the *cosine of* 17° (abbreviated, cos 17°). Hence, in the figure,

$$\frac{87.2}{x} \approx .956 \qquad x \approx \frac{87.2}{.956} \approx 91.2.$$

(Explain.)

The Tangent Function

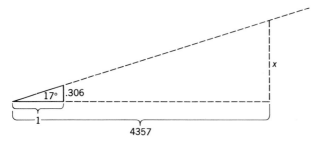

In a 17° right triangle the ratio of the side opposite the 17° angle to the side adjacent to the 17° angle is approximately .306. This ratio is called the *tangent* of the acute angle.

Tangent of 17° $\approx$ .306 $\approx$ tan 17°.

Hence, in the figure,

$$\frac{x}{4357} \approx .306 \qquad \text{and} \qquad x \approx 1333.$$

We have used the symbol $\approx$ to indicate that the numbers used are only approximations of actual distances. In the remainder of the chapter we shall simply use the equals sign.

The diagram below summarizes the definitions of the sine, cosine, and tangent functions.

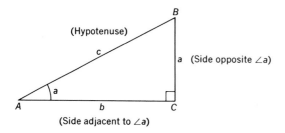

Let $\angle A$ be an acute angle of the right triangle ABC with sides of lengths indicated. Then,

$$\sin A = \frac{a}{c} = \frac{\text{opp.}}{\text{hyp.}}, \qquad \cos A = \frac{b}{c} = \frac{\text{adj.}}{\text{hyp.}}, \qquad \tan A = \frac{a}{b} = \frac{\text{opp.}}{\text{adj.}}.$$

Explain the use of the abbreviations opp. and adj. in these definitions.

The accompanying table lists three-place approximations for these functions for angles whose measures are a whole number of degrees from 0° to 90°. Explain the entries for 0° and 90°.

Angle	Sin	Cos	Tan	Angle	Sin	Cos	Tan
0°	.000	1.000	.000	45°	.707	.707	1.000
1°	.017	1.000	.017	46°	.719	.695	1.036
2°	.035	.999	.035	47°	.731	.682	1.072
3°	.052	.999	.052	48°	.743	.669	1.111
4°	.070	.998	.070	49°	.755	.656	1.150
5°	.087	.996	.087	50°	.766	.643	1.192
6°	.105	.995	.105	51°	.777	.629	1.235
7°	.122	.993	.123	52°	.788	.616	1.280
8°	.139	.990	.141	53°	.799	.602	1.327
9°	.156	.988	.158	54°	.809	.588	1.376
10°	.174	.985	.176	55°	.819	.574	1.428
11°	.191	.982	.194	56°	.829	.559	1.483
12°	.208	.978	.213	57°	.839	.545	1.540
13°	.225	.974	.231	58°	.848	.530	1.600
14°	.242	.970	.249	59°	.857	.515	1.664
15°	.259	.966	.268	60°	.866	.500	1.732
16°	.276	.961	.287	61°	.875	.485	1.804
17°	.292	.956	.306	62°	.883	.469	1.881
18°	.309	.951	.325	63°	.891	.454	1.963
19°	.326	.946	.344	64°	.899	.438	2.050
20°	.342	.940	.364	65°	.906	.423	2.145
21°	.358	.934	.384	66°	.914	.407	2.246
22°	.375	.927	.404	67°	.921	.391	2.356
23°	.391	.921	.424	68°	.927	.375	2.475
24°	.407	.914	.445	69°	.934	.358	2.605
25°	.423	.906	.466	70°	.940	.342	2.747
26°	.438	.899	.488	71°	.946	.326	2.904
27°	.454	.891	.510	72°	.951	.309	3.078
28°	.469	.883	.532	73°	.956	.292	3.271
29°	.485	.875	.554	74°	.961	.276	3.487
30°	.500	.866	.577	75°	.966	.259	3.732
31°	.515	.857	.601	76°	.970	.242	4.011
32°	.530	.848	.625	77°	.974	.225	4.332
33°	.545	.839	.649	78°	.978	.208	4.705
34°	.559	.829	.675	79°	.982	.191	5.145
35°	.574	.819	.700	80°	.985	.174	5.671
36°	.588	.809	.727	81°	.988	.156	6.314
37°	.602	.799	.754	82°	.990	.139	7.115
38°	.616	.788	.781	83°	.993	.122	8.144
39°	.629	.777	.810	84°	.995	.105	9.514
40°	.643	.766	.839	85°	.996	.087	11.430
41°	.656	.755	.869	86°	.998	.070	14.301
42°	.669	.743	.900	87°	.999	.052	19.081
43°	.682	.731	.933	88°	.999	.035	28.636
44°	.695	.719	.966	89°	1.000	.017	57.290
45°	.707	.707	1.000	90°	1.000	.000	—

The examples below illustrate the use of the table.

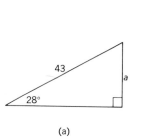

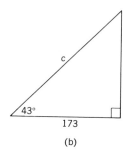

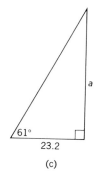

(a)

$$\frac{a}{43} = \sin 28 = .469$$

$$a = 20.2$$

(b)

$$\frac{173}{c} = \cos 43 = .731$$

$$c = \frac{173}{.731} = 237$$

(c)

$$\frac{a}{23.2} = \tan 61 = 1.804$$

$$a = 23.2(1.804) = 41.9$$

8 Circles, Spheres, Cylinders, and Cones

8–1 Circles and Spheres

A *circle* with *center* P and *radius* $\overline{PA}$ is a simple, closed, plane curve such that each point of the circle determines with P a segment congruent to $\overline{PA}$.

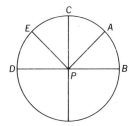

$\overline{PA} \cong \overline{PB} \cong \overline{PO} \cong \overline{PC} \cong \overline{PE}$
Circle with center P and radius $\overline{PA}$

There are infinitely many planes which contain $\overline{PA}$; hence there are infinitely many circles with P as center and $\overline{PA}$ as radius. A *sphere* with center P and radius $\overline{PA}$ consists of all the points in space on all the circles determined by P and $\overline{PA}$.

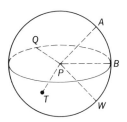

$\overline{PA} \cong \overline{PB} \cong \overline{PW} \cong \overline{PT} \cong \overline{PQ}$
Sphere with center P and radius $\overline{PA}$

Alternative definitions of a circle and sphere, using measure, are as follows:

A circle with center P and radius r consists of all points in a plane containing P at a distance r from P.

A sphere with center P and radius r consists of all points in space at a distance r from P.

Note that radius has been used in two senses: a segment and a number.

A *chord* of a circle is a segment with endpoints on the circle. A *secant* of a circle is a line which intersects the circle in two points. Clearly each chord determines a secant and each secant includes a chord. (Define chord and secant for a sphere.)

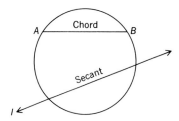

A *diameter* of a circle is a chord which contains the center of the circle. Any segment determined by the center and a point of the circle is a *radius* of the circle. (Define diameter and radius for a sphere.)

If a plane intersects a sphere in more than one point, the intersection is a circle. To see this, consider such a plane α which does not contain the center P of the sphere. Drop a perpendicular from P to α at P', and consider all points in α at a distance r from P, where r is the radius of the sphere. These are just the points of intersection of α and the sphere. (Why are the segments $\overline{P'X}$, $\overline{P'Z}$,

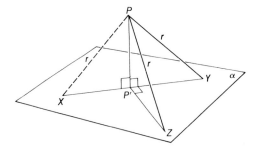

and $\overline{P'Y}$ congruent in the figure?) It follows that all points in the plane α and on the sphere lie on a circle with center P'. Clearly the intersection of a sphere and a plane containing the center of the sphere is a circle with the same center as the sphere and a radius equal to the radius of the sphere. A circle on a sphere determined by a plane containing the center of the sphere is called a *great circle*. A circle on a sphere determined by a plane not containing the center is called a *small circle*. (Why this language, "great" and "small"?)

The *interior* of a circle with center P and radius r consists of all points Y such that the distance from P to Y is less than r. The exterior of the circle consists of all points Y such that the distance from P to Y is greater than r. Define interior and exterior of a sphere.

Two circles or two spheres are congruent if they have congruent radii.

Three *concentric* circles are pictured in the figure. Define concentric circles and concentric spheres.

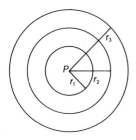

Exercises 8–1

1. Let P be the center of a circle with radius r, Q the center of a circle with radius s, and t the distance between the centers of the two circles. Draw circles for which the following are true.

   a) $r + s > t$ b) $r + t > s$ c) $r + s < t$

   d) $r + t < s$ e) $r + s = t$ f) $r + t = s$

2. The *line of centers* is the line determined by the centers of two circles. Given two circles with radii r and s, with t the distance between their centers. Complete the following theorems:

   a) The circles will intersect in two points on opposite sides of the lines of centers if...

   b) The circles will intersect in one point if...

   c) The circles will not intersect if...

3. Indicate whether each statement is true or false.

   a) A diameter of a circle is a secant of the circle.

   b) All radii of a sphere are congruent.

   c) Every diameter of a sphere is a diameter of infinitely many great circles.

   d) A radius of a circle is a also a chord of the circle.

   e) A secant of a sphere intersects the sphere in exactly one point.

f) A chord of a circle contains exactly two points of the circle.

g) A sphere and a great circle of the sphere have the same center and the same radius.

4. Which of the following statements do you think are true ?

   a) If a radius bisects a chord of a circle, then it is perpendicular to the chord.

   b) The intersection of a line with a circle may contain no points.

   c) Two circles may intersect in exactly three points.

   d) A line may intersect a circle in exactly one point.

   e) Two spheres may intersect in exactly one point.

   f) Two spheres may intersect in a circle.

   g) The secant which is the perpendicular bisector of a chord of a circle contains the center of the circle.

   h) If a line intersects a circle in one point, it intersects the circle in two points.

5. In the figure, $\overline{XY}$ is a chord of the circle with center P, and $\overline{PA}$ is a radius perpendicular to $\overline{XY}$.

   a) $\overline{PY} \cong \overline{PX}$. Why?

   b) $\triangle PQX \cong \triangle PQY$. Why?

   c) $\overline{XQ} \cong \overline{QY}$. Why?

   d) State the theorem established by (a), (b), and (c).

6. Prove that a radius which bisects a chord, not a diameter, is perpendicular to the chord. Compare this theorem with your theorem of 5(d). How are the two theorems related ?

7. Consider the three chords determined by three noncollinear points on a circle. Where will the perpendicular bisectors of these chords be concurrent ? What has this to do with constructing the circle circumscribing a given triangle ?

8. If two chords are equidistant from the center of a circle, how do their lengths compare ? If not equidistant, which is the longer ? Prove your assertions.

9. $\overline{AB}$ and $\overline{CD}$ are diameters of the circle with center P.

   a) What is true about $\overline{AD}$ and $\overline{CB}$? $\overline{AC}$ and $\overline{BD}$?

   b) State your findings as a theorem about quadrilateral $ACBD$.

   c) Prove the theorem.

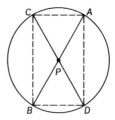

10. Explain how any two points on a sphere determine a unique great circle.
11. Study the figure and complete the following statement:

In a circle, the midpoints of all the chords congruent to one chord lie on a . . .
Verify your conjecture.

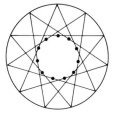

8–2 Cylinders and Cones

The surface of the solid in the figure is a cylinder. The construction of a cylinder is the same as for a parallelepiped. The bases of a cylinder lie in parallel planes and consist of closed curves and their interiors. If the bases are circular, then the cylinder is called a *circular cylinder*.

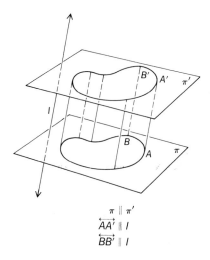

$$\pi \parallel \pi'$$
$$\overleftrightarrow{AA'} \parallel l$$
$$\overleftrightarrow{BB'} \parallel l$$

In the figure above, $\overline{BB'}$ and $\overline{AA'}$ are *elements* of the cylinder. If $l \perp \pi$ (and hence $l \perp \pi'$), then the surface is called a *right* cylinder; if l is not perpendicular to the planes, it is called an *oblique* cylinder. The segment determined by the centers of the bases of a circular cylinder is called the *axis* of the cylinder. Define *altitude* for a cylinder.

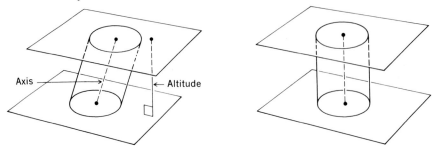

Oblique circular cylinder   Right circular cylinder

The *cone* is related to the pyramid in the same manner as the cylinder to a parallelepiped.

Formulate definitions for the concepts suggested by the accompanying figures.

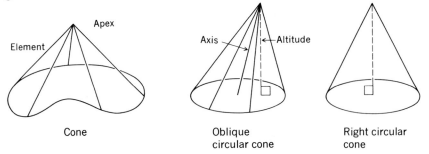

Cone   Oblique   Right circular
                              circular cone   cone

Exercises 8–2

1. Name some common objects that suggest right circular cylinders.

2. Name some common objects that suggest right circular cones.

3. Describe the space figures you could form by cutting out and folding the figures below.

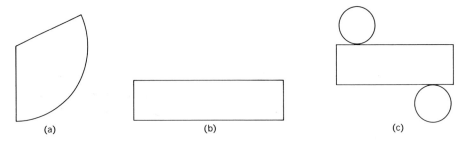

(a)   (b)   (c)

4. Make a pattern for a right circular cylinder with a base four inches in diameter and an altitude of five inches. Then construct this cylinder.

5. Draw a pattern for a right circular cone with a base five inches in diameter and an altitude of four inches. Then construct this cone.

6. Sketch the intersections of the planes with cones or cylinders.

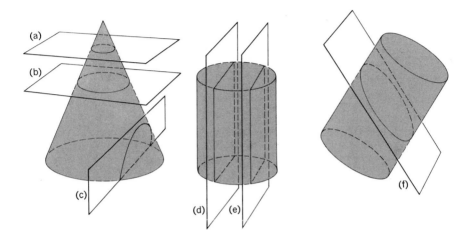

7. Describe the intersection of a cylinder and a plane parallel to a base of the cylinder.

8. Describe the intersection of a cone and a plane parallel to the base.

9. In what kind of cylinder is an element also an altitude of the cylinder?

10. A right section of a cylinder is the intersection of the cylinder and a plane perpendicular to an element of the cylinder.

   a) What is the right section of a right cylinder?

   b) What is the right section of an oblique circular cylinder?

8–3 Tangent Lines and Planes

A line is *tangent* to a circle if it is coplanar with the circle and intersects the circle in exactly one point. In (a), line *l* is tangent to the circle at point *P*. The point *P* is called the *point of tangency*. In (b), the circle is tangent to plane α at *S*. In (c), the line *m* is tangent to the sphere at point *R*. And in (d), the sphere is tangent to the plane π at point T. Formulate definitions for a circle tangent to a plane, a line tangent to a sphere, and a sphere tangent to a plane.

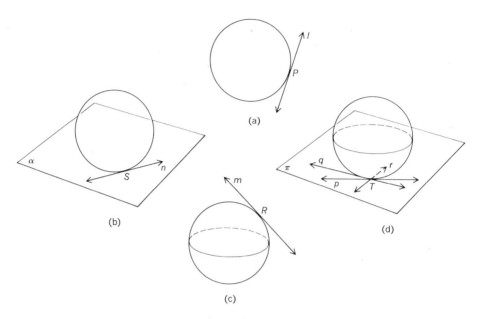

(a)

(b)

(c)

(d)

Two circles are tangent if they are tangent to the same line at the same point. Two tangent coplanar circles are said to be tangent *externally* if their centers are on "opposite" sides of the common tangent; and tangent *internally* if their centers are on the same side of the common tangent.

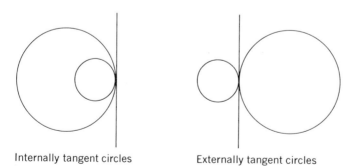

Internally tangent circles   Externally tangent circles

Consider a line perpendicular to a radius of a circle at its outer end (in the figure, $l \perp \overline{OP}$ at P). Let T be any point on l different from P. Is $\overline{OT}$ greater than, congruent to, or less than $\overline{OP}$? Since $\overline{OT}$ is the hypotenuse of a right triangle, we must have $\overline{OT} > \overline{OP}$ for any point T on l different from P. This implies that P is the only point of l which is on the circle, therefore, l is tangent to the circle at P. We state this result as a theorem.

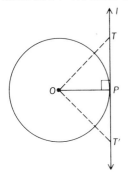

Theorem. A line perpendicular to a radius of a circle at its outer end is tangent to the circle.

If we use the theorem,

> The shortest distance from a point P to a plane π is the length of the perpendicular from P to π,

we obtain a similar theorem for a sphere and a plane. State this theorem. The converses of the theorems stated above are also true. State these theorems. We shall use these theorems in the following exercises.

Exercises 8–3

1. A wire hoop and a pencil can be used as a physical model for a line tangent to a circle. Name physical models for

   a) A line tangent to a sphere.

   b) A circle tangent to a plane.

   c) A sphere tangent to a plane.

2. Consider a pair of externally tangent circles at P, with centers O and O'.

   a) What is true about $\overline{OP}$ and l? $\overline{O'P}$ and l? Why?

   b) What is true about points O, O', and P? Why?

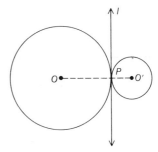

3. Consider a pair of circles internally tangent at point P, with centers O and O'.

   a) What is true about $\overline{OP}$ and l? Why?

   b) What is true about $\overline{O'P}$ and l? Why?

   c) What is true about points O, O', and P? Why?

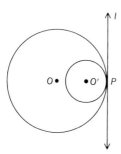

4. Complete the following statement:

   The line of centers of two tangent circles . . .

5. Consider a circle and a coplanar point P exterior to the circle. If l is a line on P, then what possibilities are there for the intersection of l and the circle? How many tangents are there to the circle through P?

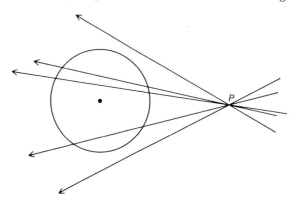

6. Draw a circle with center O. Draw two radii, $\overline{OT}_1$ and $\overline{OT}_2$. Bisect angle T_1OT_2. Construct the tangents to the circle at T_1 and T_2. (a) When will the tangents be parallel? (b) What appears to be true about the three lines (angle bisector and the tangents)? (c) Make a conjecture.

7. A circular disk consists of a circle and its interior. An instrument for locating the center of circular disks, called a center square, is made from two pieces of

metal, CQ and DQ, and a third piece PQ one edge of which bisects $\angle CQD$. Explain how the center of a disk may be found with this instrument.

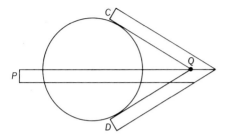

8. A line tangent to two circles is called a *common tangent*. Draw pictures of two circles with (a) no common tangents, (b) exactly one common tangent, (c) exactly two common tangents, (d) exactly three common tangents, (e) exactly four common tangents.

9. Consider a pair of concentric circles. What is true about all the chords of the greater circle which are tangent to the smaller circle? For each chord what is the relation of the point of tangency to the chord?

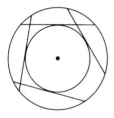

10. Lines m and n are perpendicular at P. Where are all the centers of the circles that are tangent to m at point P?

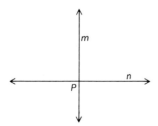

11. Consider two distinct points. Construct several circles containing these two points. (a) How many circles contain a pair of points? (b) Make a conjecture.

12. The triangle is a right triangle and the dashed curves are semicircles. What is the area of the rectangle?

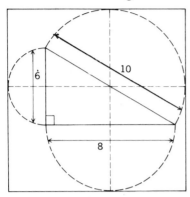

13. A sphere with center O is tangent to plane π at P. Points A, B, and C lie in π. What is the relationship of $\overleftrightarrow{OP}$ to $\overleftrightarrow{PA}$, $\overleftrightarrow{PB}$, and $\overleftrightarrow{PC}$? Explain.

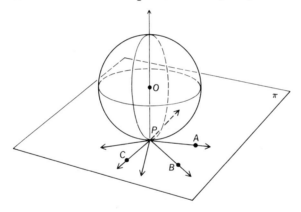

8–4 Arcs and Angles

Use the figures below in a review of definitions. The symbol $\overparen{ACB}$ indicates the arc from A through C to B. Define arc, chord, central angle, and inscribed angle.

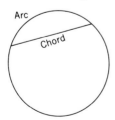

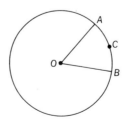

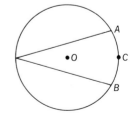

Arc and chord,
subtending each other

Central angle,
subtending arc $\overparen{ACB}$

Inscribed angle,
subtending arc $\overparen{ACB}$

Since two points of a circle determine two arcs, we need to distinguish between the two. This is done by identifying a third point of the arc as we have already done, or by using the terms major and minor arc (see the figure). Define major and minor arc. Note that $\overline{AB}$ must not be a diameter. (Why?)

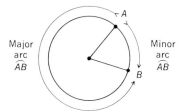

We assign to arcs of a circle the degree measure of the central angles they subtend. Thus in the figure $m\,\widehat{ED} = 10°$, since $m \angle EOD = 10°$. The measure of $\widehat{CAD}$ is 180° or half the arc measure of the complete circle. Arc $\widehat{CAD}$ is called a semicircle. We define congruent arcs as arcs in congruent circles which have the same measure.

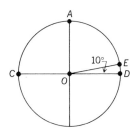

Exercises 8–4

1. In the figure, A and B are the endpoints of a diameter.
   a) Name two semicircles.
   b) Name four minor arcs.
   c) Name four major arcs.

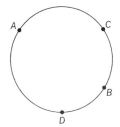

2. In the figure, O is the center of the circle. Given that $m \angle POQ = 75°$, find $m \ \overarc{PQ}, m \ \overarc{PR}, m \ \overarc{PQR}, m \ \overarc{PRQ}$.

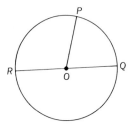

3. In the figure $\overline{AC}$ and $\overline{BD}$ are diameters. If $m \angle ACD$ is 40°, what is the measure of each of the minor arcs shown?

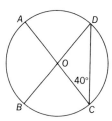

4. Central angle ABC has measure 60°.

   a) Explain why $\triangle ABC$ is equilateral.

   b) What is the measure of $\overarc{ADC}$?

   c) How many nonoverlapping arcs with the same measure as $\overarc{ADC}$ can be cut on the circle?

   d) Explain how to construct a regular hexagon.

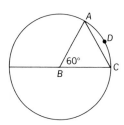

5. Explain how to trisect a right angle.

6. Using a compass, inscribe a regular pentagon in a circle by trial and error.

7. Draw a circle and $\triangle ABC$ with vertices on the circle. Let P be any point on the circle. Drop perpendiculars from P to $\overleftrightarrow{AB}$, $\overleftrightarrow{BC}$, and $\overleftrightarrow{CA}$, and call the feet

of the perpendiculars X, Y, Z, respectively. What is true about the points X, Y, Z? Try several different points on the circle. Is the same thing always true about points X, Y, Z?

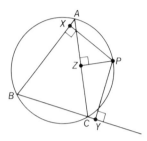

The result in Exercise 7 is due to Robert Simson, an English mathematician. The line on points X, Y, Z is called the *Simson line* associated with point P.

8. Angle BAC is an inscribed angle and $\overline{BA}$ is a diameter. Draw the radius $\overline{PC}$.

 a) Why is $\angle A \cong \angle C$?

 b) What is the relationship between $\angle A$ and $\angle BPC$?

 c) Compare the measures of $\angle A$ and $\overset{\frown}{BC}$.

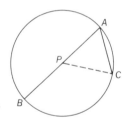

9. Angle BAC is an inscribed angle with diameter $\overline{AD}$ in the interior of the angle.

 a) Using Exercise 8, compare the measures of $\angle BAD$ and $\overset{\frown}{BD}$, $\angle CAD$ and $\overset{\frown}{DC}$.

 b) Compare the measures of $\angle BAC$ and $\overset{\frown}{BDC}$.

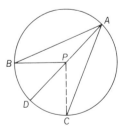

10. Angle *BAC* is an inscribed angle with the diameter of the circle from *A* exterior to the angle.

a) Using Exercise 8, compare the measures of $\angle BAD$ and $\overset{\frown}{BCD}$, $\angle CAD$ and $\overset{\frown}{CD}$.

b) Use part (a) to compare the measures of $\angle BAC$ and $\overset{\frown}{BC}$.

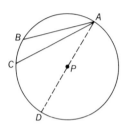

11. Using Exercises 8, 9, and 10, state a theorem about measures of inscribed angles and subtended arcs.

12. Why is an angle inscribed in a semicircle a right angle?

13. Verify that the sum of the measures of the opposite angles of any quadrilateral inscribed in a circle is two right angles. How can you describe for a given quadrilateral whether or not it is possible to circumscribe it with a circle, i.e., draw a circle through its vertices?

14. Explain how to inscribe a square in a circle.

15. Explain the construction of the tangents $\overline{PA}$ and $\overline{PB}$ from a point *P* in the exterior of the circle with center *O*. (See figure on the left below.)

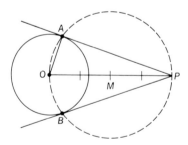
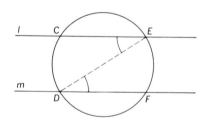

16. In the figure above, on the right, $l \parallel m$. Using the theorem of Exercise 11 and the alternate interior angle theorem, prove that $\overset{\frown}{CD} \cong \overset{\frown}{EF}$. Complete the following statement: Parallel secants . . .

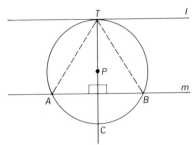

Consider a tangent l and a secant m of a circle with center P such that $m \parallel l$. If a diameter $\overline{CT}$ is drawn to the point of tangency T, then $\overline{CT} \perp m$ and $\overline{CT}$ bisects the chord $\overline{AB}$ determined by m. (Why?) It is easy to show that $\angle A \cong \angle B$, and hence $\widehat{TA} \cong \widehat{TB}$. (Verify this.) We have the following theorem:

A tangent and a chord parallel to the tangent intercept two congruent arcs.

The preceding theorem is the wedge that we need to establish results pertaining to measures of the angles formed by tangents and secants. The following exercises are designed to help you obtain these results.

Exercises 8–5

1. Consider a tangent l to a circle at A, and a chord $\overline{AC}$. Take line m on C parallel to l.

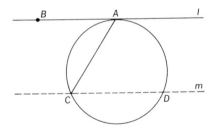

a) If $\overline{AC}$ is a diameter, what is the measure of $\angle BAC$? Explain.

b) If $\overline{AC}$ is not a diameter, then m will intersect the circle in a second point D. Explain.

c) $\angle BAC \cong \angle ACD$. Why?

d) $m\angle ACD = \frac{1}{2}m\widehat{AD}$. Why?

e) $\widehat{AD} \cong \widehat{AC}$. Why?

f) How is the measure of $\angle BAC$ related to the measure of $\overset{\frown}{AC}$?

g) Write a theorem.

2. Using the figure and statements (a), (b), (c), and (d), find a relation between $\angle BAC$ and the arcs $\overset{\frown}{BD}$ and $\overset{\frown}{BC}$ intercepted by the tangent and secant. Express your results in the form of a theorem.

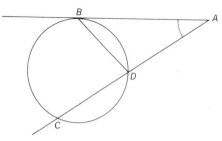

a) $\angle BDC = \angle DBA + \angle BAC$. (Why?)

b) $m\angle BDC = \frac{1}{2}m\overset{\frown}{BC}$. (Why?)

c) $m\angle DBA = \frac{1}{2}m\overset{\frown}{BD}$. (Why?)

d) $m\angle BAC = ?$

3. Use the figure below to find a relation between angle BAC and arcs $\overset{\frown}{FE}$ and $\overset{\frown}{BC}$. $\overset{\leftrightarrow}{AG}$ is tangent to the circle at G. Express your results in the form of a theorem.

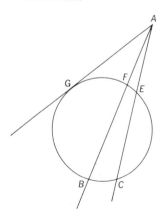

4. P is the center of the circle, $\overline{AB}$ a diameter. $\overset{\leftrightarrow}{DA}$ and $\overset{\leftrightarrow}{DC}$ are tangents. The measures of $\overset{\frown}{AE}$, $\overset{\frown}{EG}$, and $\overset{\frown}{AC}$, are shown. Find the measures of the angles numbered 1–9 in the following figure.

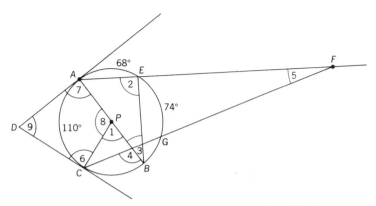

5. In order, find the measures of the angles numbered 1–4.

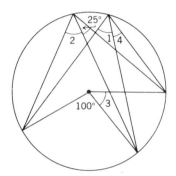

8–5 Regular Polygons and Polyhedrons

A polygon is *inscribed* in a circle if each of its vertices is a point on the circle; we say also that the circle *circumscribes* the polygon. A polygon circumscribes a circle if each of its sides is tangent to the circle; we also say that the circle is inscribed in the polygon.

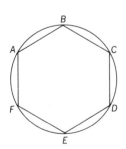

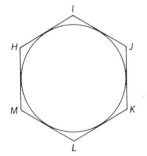

Hexagon inscribed in circle
and circle circumscribing hexagon.

Hexagon circumscribing circle
and circle inscribed in hexagon.

A *regular polygon* is a convex polygon which has all sides congruent and all angles congruent. It is both equilateral and equiangular. A *regular polyhedron* is the surface of a convex solid whose faces are congruent regular polygons. Examples of regular polyhedrons are a cube whose faces are squares and a regular tetrahedron whose faces are congruent equilateral triangles.

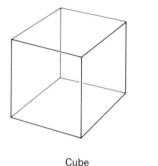
Cube

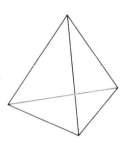
Regular tetrahedron

There is a convenient relationship between regular polygons and circles. We explore this relationship in the following exercises.

Exercises 8–6

1. Which of these simple polygons with all sides congruent can be inscribed in a circle?

2. What special quadrilateral is equilateral but not regular?

3. Sketch a polygon which (a) has all its sides congruent but is not regular, (b) has all its angles congruent but is not regular, (c) has all its sides congruent and adjacent sides forming right angles but is not regular.

4. Describe a technique for inscribing a regular polygon in a circle.

5. Explain how to inscribe a regular pentagon in a circle using only a compass to locate points on the circle.

6. Draw a triangle and circumscribe the triangle.

7. Draw a quadrilateral and circumscribe the quadrilateral.

   a) Is it possible to circumscribe any quadrilateral?

   b) Make a conjecture.

8. Draw a triangle and construct the inscribed circle. (The center of this circle is called the *incenter* of the triangle.)

9. The *Gergonne point* of a triangle is the point of concurrency of the lines joining each vertex with point of contact of inscribed circle with opposite side. In what type of triangle is the incenter equal to the Gergonne point?

10. A *cyclic quadrilateral* has opposite angles that are supplementary.

    a) Verify that any quadrilateral that can be inscribed in a circle is cyclic. (Use the inscribed angle theorem.)

    b) Can all cyclic quadrilaterals be inscribed in circles?

    c) If so, explain how.

11. Consider the figure.

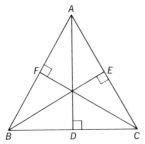

    a) Verify that F and D lie on a circle with $\overline{AC}$ as a diameter. Hence the quadrilateral $FDCA$ is cyclic.

    b) Name two other cyclic quadrilaterals in the figure.

12. Complete the following table.

Regular Polygons

Number of sides	Sums of measure of angles	Measure of interior angle	Measure of exterior angle
3			
4			
5			
6			
7			
10			
15			
27			
$\vdots$			
n			

13. How many sides has a regular polygon if the degree measure of one of the angles is (a) $128\frac{4}{7}$? (b) 140? (c) 144? (d) 160?

14. How many sides has a regular polygon if the degree measure of an exterior angle is (a) 72? (b) 45? (c) 36? (d) $17\frac{1}{7}$?

15. Choose seven points on a circle and label them in order A, B, C, D, E, F, and G. Draw the polygon $ACEGBDFA$.

   a) What figure do you get?

   b) Would the same method of skipping vertices work for an octagon?

   c) Explain how to draw an inscribed eight-pointed star.

16. How many regular polyhedrons with square faces can be constructed? Explain.

17. Cut out of stiff cardboard several congruent equilateral triangles with flanges, as above. Using rubber bands to hold the sides of the triangles together, how many regular polyhedrons can you construct?

18. Using congruent regular pentagons with flanges, how many regular polyhedrons can you construct?

19. Is it possible to use regular polygons other than triangles, squares, and pentagons for faces of a regular polyhedron? Explain. How many regular polyhedrons are there?

20. Consider any polyhedron, regular or not. Additional polyhedrons can be obtained by slicing corners from polyhedrons, as illustrated.

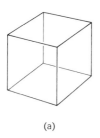
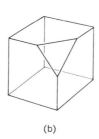
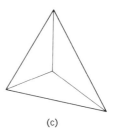
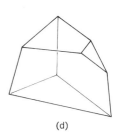

(a)   (b)   (c)   (d)

Complete the following table for the polyhedrons pictured at the bottom of page 192.

Figure	Number of faces	Number of vertices	Number of edges
(a)	6		
(b)		10	
(c)			6
(d)			

A relationship known as Euler's formula relates the three entries in this table. Can you guess the formula by using the data you have listed?

8–6 Circumference and Area of a Circle

The *circumference* of a circle is, roughly speaking, the length of the circle or the distance around it. It is analogous to the perimeter of a polygon. Use a string and a measuring stick or a tape measure to find the approximate circumference and diameter of several circular objects. Make a table in the following form.

Circumference (C)	Diameter (D)	C/D
$\vdots$	$\vdots$	$\vdots$

Express the ratio C/D in the third column in decimal form correct to two places. What appears to be true?

We denote the ratio C/D, which is the same for all circles, by the Greek letter π (pi). There are techniques for approximating π. Some tables give π correct to as many as 100,000 decimal places. We list below approximations which are accurate enough for the problems we wish to solve.

$$\pi \approx 3.1416 \approx 3.142 \approx 3.14 \approx \tfrac{22}{7}.$$

Consider a regular pentagon inscribed in a circle with center P. The point P is also called the center of the polygon. The segment $\overline{PQ}$ is called an *apothem* of the polygon, and its length is denoted by a. (Define apothem for a regular polygon.) If we denote the length of a side of the pentagon by e, then the perimeter is $5e$ and the area is $5 \cdot \frac{1}{2}ae$. (Why?) In general, if a regular polygon of n sides is inscribed in a circle, the perimeter of the polygon is ne and the area is $n \cdot \frac{1}{2}ae$.

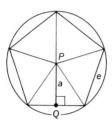

Now consider a sequence of regular polygons of 6, 12, and 24 sides inscribed in the same circle. Denote the apothems by a_6, a_{12}, a_{24}, the length of the sides by e_6, e_{12}, e_{24}, and the perimeters and areas of the polygons by P_6, P_{12}, P_{24} and A_6, A_{12}, A_{24}, respectively. Compare the three apothems with the radius, the three perimeters with the circumference, and the three areas with the area of the circle. As the number of sides increases the apothem approaches the radius of the circle, the perimeter of the polygon is a good approximation for the circumference of the circle, and the area of the polygon is very nearly equal to the area of the circle. Thus when n is very large

$$P_n = ne \approx C = 2\pi r,$$

and

$$A_n = n \cdot \tfrac{1}{2}ae = \tfrac{1}{2}(n \cdot e)a \approx \tfrac{1}{2}(2\pi r)r = \pi r^2.$$

This is a persuasive argument that the area of the circle is exactly πr^2.

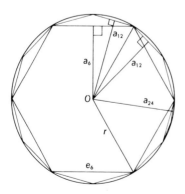

Exercises 8–7

1. A bicycle wheel has a diameter of 25 inches.

   a) How far does the bicycle travel with each revolution of the wheel?

   b) How many revolutions will it make per mile?

2. Find a relationship between the apothem a, the length of a side e, and the radius of the circle in the case of a regular inscribed polygon.

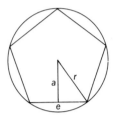

3. The side of a square is eight inches long. Find the circumference of its inscribed circle; of its circumscribed circle.

4. Consider a circle with an inscribed regular polygon of 18 sides. Cut out the 18 triangles determined by the vertices and the center of the polygon, and assemble them to form a parallelogram.

   a) What is the approximate height of the parallelogram?

   b) What is the approximate length of the base of the parallelogram?

   c) How does the area of the parallelogram compare with the area of the circle?

5. What is the area of the annulus (the shaded region in the figure on the left, below)?

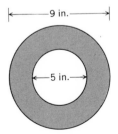

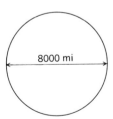

6. Suppose that the earth is a sphere and that a string is wrapped around it at the equator. (See the figure on the right, above.) Now suppose that the string is lengthened to make a circle standing out six inches from the equator. How much longer must the string be? (First, guess how much.)

7. Point O is the center of the large circle on the left. The small arcs are semi-circles.

   a) What is the area of the shaded sector? Explain.

   b) Show how to draw one straight line that bisects both areas.

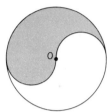 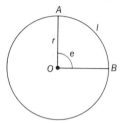

8. The *radian* measure of a central angle is the ratio of the length of the subtended arc to the radius.

   a) What is the sum of the angles about the center of a circle (or about a point) in radian measure?

   b) What is the radian measure of a straight angle? A right angle?

   c) What is the radian measure of a 60° angle? A 150° angle? A 45° angle?

   d) What is the approximate equivalent in degree measure of one radian?

9. A diagonal of a polygon is a segment determined by two nonconsecutive vertices. Complete the table below.

Polygon	Number of diagonals
Quadrilateral	2
Pentagon	5
Hexagon	
7-gon	
8-gon	
9-gon	

   a) Do you see a pattern? Explain.

   b) Without counting, how many diagonals for a 10-gon? An 11-gon? A 20-gon?

   c) How many diagonals for an n-gon?

8–7   Volume and Surface Area of Cylinders and Cones

The formulas for the volume and surface area of cylinders and cones can be obtained from the corresponding formulas for prisms and pyramids. We use the same techniques used earlier in finding the formula for the area of a circle.

Consider a prism with regular bases inscribed in a circular cylinder. We know that the volume of the prism is the area of the base times the altitude. How does the volume of the prism compare with the volume of the cylinder? Now consider a sequence of such prisms with regular polygonal bases and a greater and greater number of sides. Explain why the volume of the circular cylinder is

$$V = \text{area of base} \cdot \text{height} = \pi \cdot r^2 \cdot h.$$

The lateral surface area of the prism is the product of the perimeter of the base times the height. Write expressions for the lateral surface area and total surface area of a cylinder.

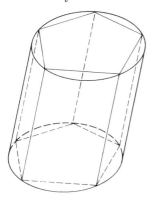

Exercises 8–8

1. Using sand (or other materials), compare the volume of a right circular cone and a right circular cylinder which have equal base radii and equal heights.

2. The two cylinders are congruent. Compare the volume of the inscribed cone on the left with the volume of the two cones on the right.

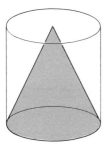 

3. In the figure on the left, a right pyramid is inscribed in a right circular cone.

a) What is the volume of the pyramid?

b) Explain why the volume of the cone is $\frac{1}{3} \cdot \pi \cdot r^2 \cdot h$.

c) Write an expression for the lateral surface area of the pyramid in terms of the perimeter of the base and the altitude a of a face. [*Hint:* Consider a face of the pyramid.]

d) Explain why the lateral surface area of the cone is $\frac{1}{2}$ the circumference times the slant height or $\pi \cdot r \cdot s$. Explain also why $s^2 = r^2 + h^2$.

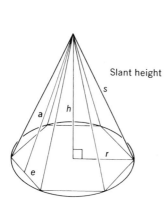

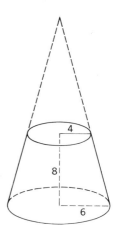

4. A frustum of a cone has altitude of 8, and the radii of its upper and lower bases are 4 and 6. What is its volume?

5. Members of a city council inspected plans for a new cylindrical water tank. They decided it should be twice as large as planned. They instructed the contractor to double the diameter but to retain the height. Criticize their decision.

8–8 Volume and Surface Area of a Sphere

Consider a sphere of radius r. How many disks of radius r do you think are needed to cover the sphere? Using sticky paper cover the sphere using disks of radius r. Make a conjecture about the surface area of a sphere of radius r.

Now consider a sphere of radius r and a right circular cylinder of radius r and height $2r$. Using water, sand, or some other material compare the volumes of the sphere and cylinder. Make a conjecture about the volume of a sphere of radius r.

If you perform the two experiments carefully, you may guess that the surface area of a sphere is four times the area of a great circle and that the volume of a sphere of radius r is $\frac{2}{3}$ the volume of a right circular cylinder with radius r and height $2r$. Therefore we have the following results:

Surface area of a sphere of radius $r = 4\pi r^2$,
Volume of a sphere of radius $r = \frac{4}{3}\pi r^3$.

Exercises 8–9

1. As shown in the figure, a sort of "pyramid" has been cut out of the ball, its vertex at O, the center of the sphere.

   a) What is the base of the "pyramid"?

   b) What is the altitude of the "pyramid"?

   c) What is the approximate volume of the "pyramid"?

   Now think of the surface of the sphere divided into small patches of area.

   d) What is the total surface area?

   e) Now explain why the volume of a sphere is $\frac{1}{3} \cdot r \cdot (4\pi r^2) = \frac{4}{3}\pi r^3$.

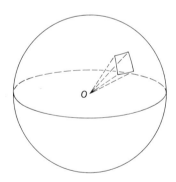 

2. Find the surface area and volume of a sphere whose radius is 4.
3. One sphere has twice the radius of another. How do their volumes compare? How do their surface areas compare?
4. What is the diameter of a sphere if its volume is equal to its surface area?
5. About three-fourths of the surface of the earth is covered with water. How many millions of square miles of its surface are land? (Use 8000 mi as the diameter, and $\pi = 3.14$.)

6. A Hamiltonian Circuit (named after the great Irish mathematician Sir W. R. Hamilton, 1805–1865) is a closed polygonal path along some of the edges of a polyhedron and *passing through each vertex once.*

A Hamiltonian Circuit is shown for a tetrahedron. Find Hamiltonian Circuits for the others.

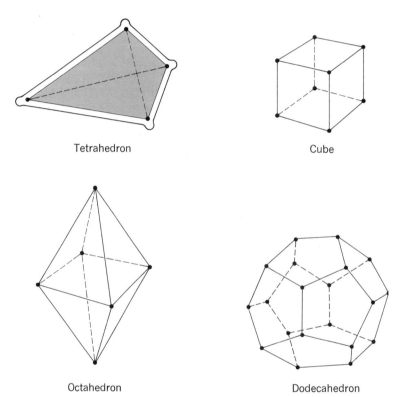

| Tetrahedron | Cube |

| Octahedron | Dodecahedron |

8–9 Elliptic Geometry

In Chapter 4, we discussed a second geometry (hyperbolic) in which the sum of the angles of a triangle is less than 180°. There is a third geometry known as *elliptic geometry* in which the betweenness relations we accept in both Euclidean and hyperbolic geometry do not hold. In this geometry there are no parallel lines and every pair of distinct lines intersects in *two* points. In elliptic geometry the exterior angle theorem is false, that is, an exterior angle of a triangle can be less than or equal to a remote interior angle. An illustration of this is spherical geometry in which *lines* are great circles on a sphere, and *segments* are arcs of great circles. A spherical *triangle* consists of three points not all on the same great circle and the three arcs of great circles containing these points. The measure of a spherical *angle* is the measure of the plane angle formed by the tangents to the great circles at the vertex of the angle. In elliptic geometry every triangle has an angle sum greater than 180°, and the larger the triangle the larger the angle sum. We picture below a spherical triangle whose angle sum is 300°. Note that an exterior angle of 90° is less than the remote interior angle of 120°.

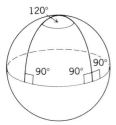

In elliptic geometry the SAS, ASA, and SSS theorems of triangle congruence hold, but the AAS theorem does not.

9   Geometric Figures as Sets of Points and Lines

9–1 Introduction

It is rather convenient to think of the common geometric figures as sets of points. This way of thinking has several advantages. It enables us to employ much of the rich, descriptive language of set theory; it simplifies the task of formulating precise definitions; and it lends itself naturally to algebraic treatments of geometric problems when coordinate systems are introduced. Hence, in this chapter we shall think of lines, circles, and triangles as sets of points. It is not essential that one think in this way. Some geometers prefer to think otherwise. For example, they visualize a line as a *support* for points. Then the points do not form the line. Rather the points are *on* the line as beads are on a string. The points of a plane may be viewed not as elements of the plane but rather in the role of grains of sand resting upon a table top. These physical analogs are clumsy, but it should be noted that thinking of lines as sets of points is not an essential feature of geometry. Of course, not all geometric configurations are point sets. A geometer may consider a set of concentric circles, or the set of all tangent lines to a given curve, or the set of all polygons that can be inscribed in a circle.

We do not define lines or planes, but we list below several "intuitively obvious" properties that lines and planes, as sets of points, should have. When we treat geometry axiomatically, these intuitively obvious properties will be replaced by a collection of assumptions, but for the moment we think of our geometry as an idealized description of the real world.

Properties

1. *Every line contains infinitely many points, but no line contains all the points of a plane.*

2. *For each pair of points A, B in a line, there is a point R of the line between A and B; a point S in the line such that B is between A and S; and a point T in the line such that A is between B and T.*

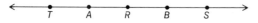

It is worth remarking that these properties are certainly idealizations. In our experience with the physical world we never encounter (knowingly) sets of infinitely many physical objects. Nor, in the physical world, can we conceive of objects so small that Property 2 holds for them.

3. *One and only one line contains a given pair of points.*

4. *The intersection of two distinct lines is empty or contains exactly one point.*

5. *If two points of a line lie in a plane, then all points of the line lie in the plane.*

   Properties 3, 4, and 5 reflect our intuitive notion of *straightness* for lines and *flatness* for planes.

6. *Each point on a line separates the line into two disjoint sets.*

7. *Each line separates the plane into two disjoint sets.*

8. *If A is any point on a line l and $\overline{XY}$ is any segment, then there are exactly two points B and B' on l such that*

$$\overline{AB} \cong \overline{XY}, \qquad \overline{AB'} \cong \overline{XY},$$

and A is between B and B'.

The properties that we have listed here are not all the properties that we need nor are they independent of one another. As a matter of fact it is relatively easy to prove that Property 4 follows from Property 3. It is possible to prove also that Property 6 is a consequence of other properties, but this proof is quite difficult. However, we are not concerned with formal proofs at the moment. Listing these properties provides us with a common basis for the problems of this chapter. We shall give more precise definitions of some of these terms later in the chapter.

In order to emphasize that it is possible to think of lines as sets of points we shall, in this section, represent lines by *Venn diagrams* instead of the conventional drawings. Hence, a diagram like the one below indicates that lines *l* and *m* have an intersection point.

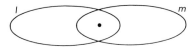

We have deliberately not marked any other points in our diagram. We agree, when points are not marked in a pictured region, that there may either be or not be points in that region; that is, referring to the diagram above, it is quite possible that all points of *l* belong to line *m*.

Exercises 9–1

In these exercises all Venn diagrams represent lines in one plane.

1. For each Venn diagram below draw ordinary "straight line" diagrams that agree with the Venn diagrams. Be able to justify your drawings by appealing to the properties listed above for lines. It may be that since some Venn diagrams give only partial information, there are several acceptable different interpretations. In that case give all possible interpretations.

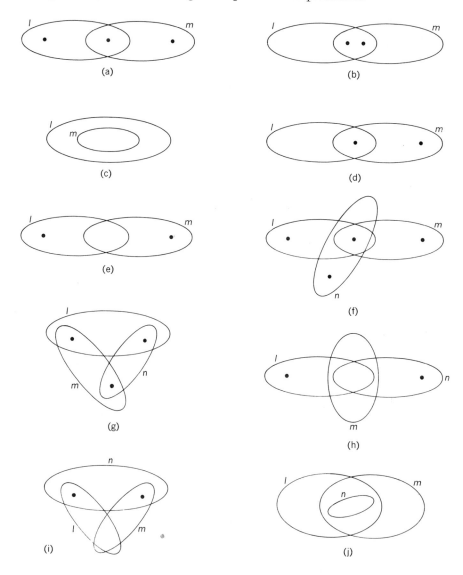

2. Explain why each diagram below presents an impossible situation if the Venn diagrams represent lines.

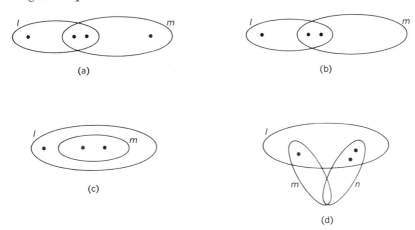

2. Explain why each diagram below presents an impossible situation if the Venn

(a)   (b)

(c)

(d)

9–2 Definitions

The definitions that we give below employ certain undefined terms like *between* and *congruent*. In a rigorous mathematical development properties of betweenness and congruence would be listed as basic axioms. These properties would then justify the use of the terms in definitions and proofs. However, our treatment is basically an intuitive one, and we trust you to accept these terms on the basis of your intuitive understanding of their meaning. When we say that B is between A and C, it is understood that A, B, and C are collinear, and we shall write (ABC) for this relation.

We shall also assume that any set of n points on a line can be named A_1, A_2, $A_3, \ldots, A_n$, so that a point is between two other points if its subscript is a number between the subscripts of the other two points.

Exercises 9–2

1. Draw a line. Mark four points A, B, C, D on the line. Write all the betweenness relations. How many are there? How many betweenness relations for five points on a line?

2. On a line there are points R, S, T, no two of which are the same. R is not between S and T. S is not between R and T. What can you conclude? Explain.

3. A, B, and C are points such that no two are the same. No one of these points is between the other two. What can you conclude? Explain.

4. A, B, and C are collinear points. A is not between B and C; B is not between A and C; and C is not between A and B. What can you conclude? Explain.

We write $(ABCD)$ for the following: (ABC), (ABD), (ACD), and (BCD).

5. If (ABC), and X is on the same line as A, B, and C, what are the possible betweenness relations for the four points, given that X is distinct from A, B, and C? Write your answer using the above notation.

6. If (ABC) and (BCD), what are the betweenness relations for the four points?

7. Indicate whether each of the following statements is true or false, and explain.

   a) If (ABC) and (BCD), then (ABD) and (ACD).

   b) If (ABC) and (ACD), then $(ABCD)$.

   c) If (ABC) and (ACD), then (ABD) and (BDC).

   d) If (ABC) and (ACD), then (BCD).

   e) If (ABC) and (ABD), then $(ABCD)$ or $(ABDC)$ or $C = D$.

8. Use Property 3 to establish Property 4.

9. Let S be a set of points, and T a subset of S. We say that T separates S into two sets A and A' if and only if

   a) $S = A \cup A' \cup T$.

   b) A, A', and T are non-empty, pairwise disjoint sets.

   c) There is a point of T between P and Q if P is any point of A, and Q is any point of A'.

   Give at least three examples of a subset separating a set. [*Note:* the separating set may consist of only one point.]

Although points, lines, and planes are undefined elements of geometry, other geometric figures are defined in terms of points and lines. Notation is not completely standardized in geometry textbooks. For example, some books designate a segment with endpoints A and B by the symbol AB; others use the notation $\overline{AB}$, as we have been doing. We shall use the notation that we consider to be most popular or convenient.

Segment. The set of points consisting of two distinct points and all points between them is called a segment. Thus if A, B are two distinct points of a plane α, the segment determined by A and B is denoted by $\overline{AB}$, and

$$\overline{AB} = \{A\} \cup \{X \in \alpha \mid (A\ X\ B)\} \cup \{B\}.$$

The points A and B are called *endpoints*, and the set $\{X \in \alpha \mid (A \ X \ B)\}$ is called the *interior* of the segment or *open segment*. The figure below indicates other sets determined by distinct points A and B, and the notations used for these sets.

Half-open segment $\overline{AB}$ Half-open segment $\overline{AB}$

$$\overline{AB} = \{A\} \cup \{X \in \alpha \mid (A \ X \ B)\} \qquad \overline{AB} = \{X \in \alpha \mid (A \ X \ B)\} \cup \{B\}$$

Some authors tolerate the condition $A = B$, and speak of *null* segments.

It is intuitively obvious that each point of a line separates the line into two sets. These sets are called *half-lines*. The point determining the half-lines does not belong to either of them.

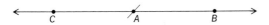

In the diagram above, one half-line determined by A contains B; and the other, C. We refer to these half-lines as B's side of A and C's side of A.

Ray. Each half-line, together with the point determining it, is called a *ray*. In the diagram above, these rays are denoted by $\overrightarrow{AC}$ and $\overrightarrow{AB}$. The point A is said to be the endpoint of each of these rays. Two rays like $\overrightarrow{AB}$ and $\overrightarrow{AC}$ on one line and with the same endpoint are *opposite rays*. Described as a set of points,

$$\overrightarrow{AB} = \{A\} \cup \{X \in \overleftrightarrow{AB} \mid X \text{ is on } B\text{'s side of } A\}.$$

It is convenient to denote the ray opposite $\overrightarrow{AB}$ by $A \swarrow B$. We read $A \swarrow B$ as "The ray with endpoint A away from B." Using these notations, how many rays can you name in the above diagram which shows three points A, B, C on a line?

In a plane, each line in the plane separates the plane into two sets, neither including the separating line. The disjoint sets are called *half-planes*. In the figure below, the points C and D are in opposite half-planes. (Define half-space.)

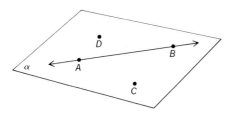

Half-planes determined by $\overleftrightarrow{AB}$.

Exercises 9–3

1. Complete the following statements:

   a) If X and Y are on opposite sides of P on a line l, then . . .

   b) If C and D are in opposite half-planes determined by $\overleftrightarrow{EF}$ in α, then . . .

   c) If R and S are in opposite half-spaces determined by plane π, then . . .

   d) If U and T are in the same half-plane determined by $\overleftrightarrow{MN}$ in π, then . . .

2. A, B, C, P are four distinct collinear points. B and A are on opposite sides of P. C is on B's side of P. What can you deduce? Explain.

3. Line l lies in plane α. Point R is on one side of l. Point S is not on R's side of l, and is not on the opposite side of l from R. What can you deduce? Explain.

4. Using the terminology we have developed, describe as simply as possible the following sets of points in a plane π. Points A and B are distinct points in π.

   a) $\{X \in \pi \mid X = A \text{ or } X = B \text{ or } (A\ X\ B)\}$

   b) $\{X \in \overleftrightarrow{AB} \mid X = A \text{ or } A \text{ is not between } X \text{ and } B\}$

   c) $\{X \in \pi \mid X = A \text{ or } X = B \text{ or } (A\ X\ B) \text{ or } (A\ B\ X)\}$

   d) $(A \not{K} B) \cup \overrightarrow{AB}$

   e) $\{A\} \cup \{X \in \pi \mid (A\ X\ B)\} \cup \{B\} \cup \{X \in \pi \mid (A\ B\ X)\}$

   f) $\{X \in \pi \mid (A\ B\ X)\} \cup \{B\}$

   g) $(B \not{K} A) \cup (A \not{K} B) \cup \overline{AB}$

5. Find the unions and intersections.

   a) $\overrightarrow{AB} \cup \overrightarrow{BA}$ b) $\overrightarrow{AB} \cap \overrightarrow{BA}$ c) $(A \not{K} B) \cap (B \not{K} A)$

   d) $\overrightarrow{AB} \cap (A \not{K} B)$ e) $(B \not{K} A) \cap AB$ f) $\overline{AB} \cup \overrightarrow{BA}$

   g) $\overrightarrow{AB} \cup (B \not{K} A)$ h) $(A \not{K} B) \cup \overrightarrow{AB}$

6. Given $(A\ C\ B)$. Find

   a) $\overrightarrow{AB} \cap \overrightarrow{CB}$ b) $(A \not{K} B) \cap (B \not{K} C)$ c) $\overrightarrow{AB} \cup \overrightarrow{CB}$

   d) $\overrightarrow{CB} \cap \overrightarrow{CA}$ e) $(A \not{K} B) \cup (B \not{K} C)$ f) $\overrightarrow{CB} \cup \overrightarrow{CA}$

7. Use the definition of a segment to prove that $\overline{AB} = \overline{BA}$. [*Hint:* If $(A\ X\ B)$, then $(B\ X\ A)$; and two sets H and K are equal if and only if $H \subset K$ and $K \subset H$.]

8. Use the definition of a ray to prove $\overrightarrow{AB} \neq \overrightarrow{BA}$.

9. Sketch figures for all possible intersections of two rays.

10. Possible intersections of two half-planes are shown on the following page. Draw figures illustrating these possibilities. Are there other types of intersections?

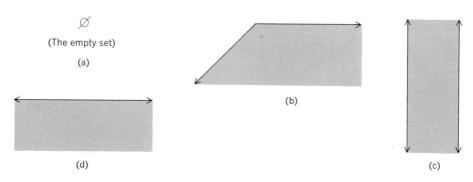

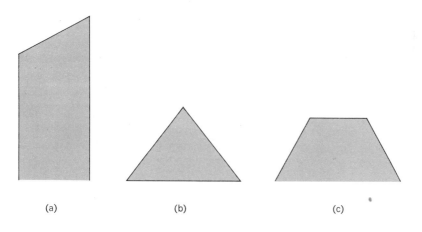

11. Three half-planes can have the sets of points of Exercise 10 as intersections.
Other possibilities are listed below. Show how these occur.

(a)   (b)   (c)

12. Use the concept of betweenness for points
to define betweenness for rays in a plane.

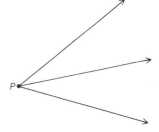

We define angle as follows:

Angle. Two rays with the same endpoint form an angle. If the rays are $\overrightarrow{BC}$
and $\overrightarrow{BA}$, the angle is denoted by $\angle ABC$ or $\angle B$. Thus

$$\angle ABC = \overrightarrow{BA} \cup \overrightarrow{BC}.$$

The point B, the common endpoint, is called the *vertex* of $\angle ABC$, and rays $\overrightarrow{BC}$ and $\overrightarrow{BA}$ are called the *sides* of the angle.

The diagram below illustrates a possible classification of angles into three types.

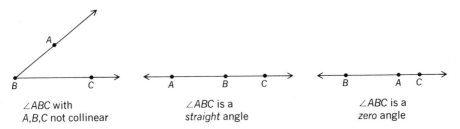

| $\angle ABC$ with A,B,C not collinear | $\angle ABC$ is a *straight* angle | $\angle ABC$ is a *zero* angle |

There is a clumsiness in defining an angle to be the *union* of its rays. With this definition a straight angle is nothing more than a line. But one line can furnish us with many different straight angles as we choose different vertex points.

It is quite clear that the concepts of a *straight angle* and a *line* are distinct concepts, and yet to define an angle as the union of its sides identifies these concepts. There are two ways to avoid embarrassment. One is to exclude straight angles from the family of angles; the other is to ignore the difficulty. We follow the latter course. Straight angles are too convenient to abandon. Interestingly, Euclid did not employ either zero angles or straight angles. We observe that the awkwardness which arises in the definition of a straight angle can be removed, but this requires a fussier treatment than we wish to present.

Triangle. Three noncollinear points determine three segments. The union of these segments is a *triangle*. If the points are named A, B, and C, the triangle is denoted by the symbol $\triangle ABC$. Thus

$$\triangle ABC = \overline{AB} \cup \overline{BC} \cup \overline{CA}.$$

The segments $\overline{AB}$, $\overline{BC}$, and $\overline{CA}$ are called *sides* of the triangle. The angles, $\angle ABC$, $\angle BCA$, $\angle CAB$, are called *angles* of the triangle. These angles are often denoted by the symbols, $\angle B$, $\angle C$, $\angle A$.

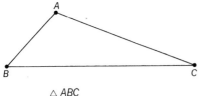

$\triangle ABC$
Sides: $\overline{BA}$, $\overline{AC}$, $\overline{CB}$
Angles: $\angle A$, $\angle B$, $\angle C$

A triangle is a special case of a *simple closed curve*. The concept of a simple closed curve is intuitively clear but too elusive for formal definition. Intuitively, a simple closed curve is any path that one can trace that returns to its initial point (closed) and does not cross itself (simple). The first two figures below are simple closed curves. The other three are not.

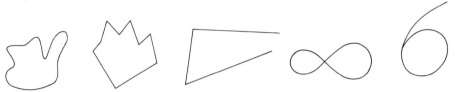

Each simple closed curve in a plane has an *interior* and an *exterior*. There is no path from an interior point to an exterior point that does not cross the curve. However, an angle also has an interior and exterior in this sense, and angles are not closed curves. Clearly, simple closed curves and angles separate the plane.

Closed curves consisting of line segments are called *polygons*. Polygon $ABC\cdots L$ is the set of points in the union of segments $\overline{AB}, \overline{BC}, \ldots, \overline{LA}$. The points $A, B, \ldots, L$ are *vertices of the polygon*; the segments $\overline{AB}, \ldots$ are *sides*; and the angles, $\angle A, \ldots$ are *angles of the polygon*.

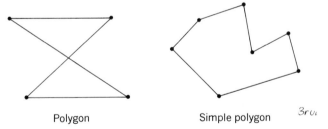

Polygon   Simple polygon   *3rᴜ₁*

We shall usually use the term polygon to mean *simple* polygon, one that does not cross itself.

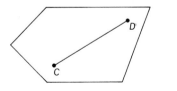 

Convex polygon   Nonconvex polygon

In the figure above, a *convex* and a *nonconvex* polygon are shown. Define a convex and nonconvex polygon.

Of all polygons the triangle seems to be the most fundamental. All polygons can be visualized as formed by gluing triangles together. The interior of any angle (excluding straight angles) and of any *convex* polygon is the intersection of half-planes. For example, the interior of $\angle ABC$ below is the intersection of two half-planes, the half-plane determined by $\overleftrightarrow{BA}$ which contains C and the half-plane determined by $\overleftrightarrow{BC}$ which contains A. These half-planes are shaded in the figure below, and the crosshatched area represents the interior of $\angle ABC$.

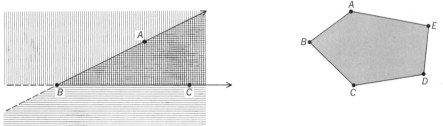

Note that the interior of the convex polygon $ABCDE$ pictured above is the intersection of five half-planes. Name them. The exterior of a polygon or any simple closed curve consists of the points in the plane not in the interior or on the curve.

One thing should be observed in our definition of angle. In some mathematics courses, trigonometry for example, angles are identified with *rotations*. That is, one considers any two rays from one point and then imagines rotating one to coincide with the other. Clearly two directions of rotation are possible, as the diagram below shows.

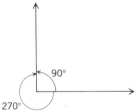

In the familiar degree notation, one rotation shown above is of 90°, the other 270°. In this case it is convenient to speak of angles of 270°. One may even speak of a 450° angle, as the figure illustrates.

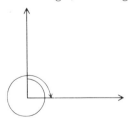

When this terminology is used, the term "angle" means much more than just a set of points. From our point of view all angles, other than straight angles, are *less* than straight angles. Their degree measures are less than 180°, for what we actually measure is the *interior* of each angle.

Other definitions are suggested in the following exercise list.

Exercises 9–4

1. In 3-space, define a dihedral angle in terms of half-planes. For a dihedral angle what is the analog of vertex of a plane angle? What types of dihedral angles are there? Define "interior of a dihedral angle."

2. Define "tetrahedron" as a set of points. Define "interior" and "exterior" for a tetrahedron.

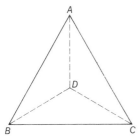

3. Draw figures showing that the intersection of two triangles can be a set of 0, 1, 2, 3, 4, 5, or 6 points. What can you say if the intersection contains more than 6 points?

4. List any interesting conclusions you can draw if you know that

   a) $\overline{AB} \subset \overline{CD}$ 　　　　b) $\overrightarrow{AB} \subset \overrightarrow{CD}$ 　　　　c) $\overleftrightarrow{AB} \subset \overleftrightarrow{CD}$

   d) $\angle ABC \subset \angle DEF$ 　　e) $\triangle ABC \subset \triangle DEF$

5. Draw a figure picturing (a) the union of the three angles of $\triangle ABC$, (b) the intersection of the three angles of $\triangle ABC$, (c) the intersection of the interiors of the three angles of $\triangle ABC$. Do you see a way to define the interior of a triangle other than the way it was done earlier in terms of half-planes?

6. A polygon is convex if its interior is a convex set. A set S is convex if for any two points A, $B \subset S$, $\overline{AB} \subset S$. (Compare with the definition you formulated earlier.)

   a) Define nonconvex.

   b) Draw several simple closed curves which are convex and several which are nonconvex.

7. Prove that half-lines, half-planes, and half-spaces are convex sets of points.

8. Define the exterior of an angle in terms of half-planes.

9. Draw a picture of a triangle and a line such that

   a) The intersection of the line and the triangle is one point,

   b) The line and the triangle intersect in exactly two points,

   c) The intersection of the line and triangle consists of more than two points.

10. Can the intersection of a line l and a triangle ABC be a single point other than A, B, or C?

11. If a line l contains a point of $\triangle ABC$ between A and B and does not contain C, what can you conclude?

12. If a line l contains a vertex of $\triangle ABC$ and an interior point of $\triangle ABC$, what can you conclude?

In Exercise 11, your conclusion should be *Pasch's Axiom*. Your conclusion of Exercise 12 should be a result which is a consequence of the Pasch Axiom. Moritz Pasch (1843–1930) was a famous geometer who recognized that this assumption was necessary to take care of an assumption unconsciously made by Euclid in proving a theorem of Book I.

9–3 Conics

To provide an example of important geometric figures which are not sets of points we give a brief discussion of point and line conics. In addition this will provide a basis for computing equations of point conics when we discuss coordinate geometry.

Roughly speaking, a *point conic* is a set of points which satisfies certain prescribed conditions, and a *line conic* is a set of lines which satisfies certain prescribed conditions. The two types of conics are related in that for each point conic there

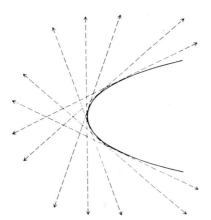

Point conic (heavy curve)
and corresponding line conic.

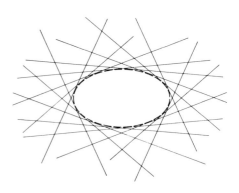

Line conic (heavy lines) and
corresponding point conic.

is a corresponding line conic and conversely. *The line conic associated with a point conic is the collection of lines tangent to the point conic.* If we have a line conic, then the associated point conic is the collection of points on one and only one of the lines of the line conic; hence it is the point conic with this set of lines as tangents. The line conic corresponding to an ellipse is called a line ellipse, etc.

There are several methods of defining a point conic. We list three below and consider other methods in the exercises.

1. *A point conic is the intersection of a plane and an infinite circular cone.* The type of conic depends on the angle made by the intersecting plane and a plane perpendicular to the axis of the cone.

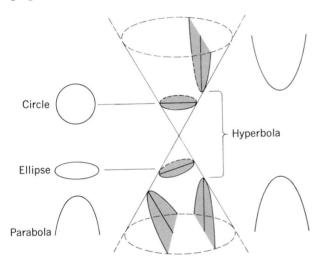

2. An *ellipse* is the set of points P in a plane such that the sum of the distances from P to two fixed points is constant. The fixed points are called the *foci* of the ellipse.

   A *parabola* is the set of points P in a plane which are equidistant from a given point and a given line. The given point is called the *focus* and the given line the *directrix*.

   A *hyperbola* is the set of points P in a plane such that the difference of its distances from two fixed points is constant.

3. *A conic is the set of points P in a plane such that the ratio of the distance from a fixed point O and a fixed line l is constant.* If we denote the constant ratio by θ, then

$$\frac{m\,\overline{OP}}{m\,\overline{PK}} = \theta.$$

If $\theta < 1$,   then the conic is an ellipse,

If $\theta = 1$ then the conic is a parabola,

If $\theta > 1$ then the conic is a hyperbola.

The circle is a special case of the ellipse.

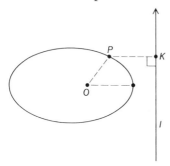

In the following exercises use a stencil to draw the various types of conics. If a stencil is not available use the conics sketched below. Graph paper will simplify much of the work in this set of exercises.

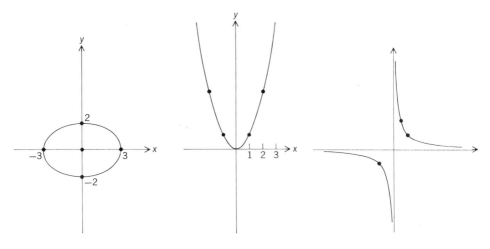

Ellipse: $\dfrac{x^2}{9} + \dfrac{y^2}{4} = 1$ Parabola: $y = x^2$ Hyperbola: $y = \dfrac{1}{x}$

Exercises 9–5

1. Consider an infinite right circular cone in which an element makes an angle of measure $45°$ with the axis. The plane α is perpendicular to the axis l at the apex V.

a) What angle must a plane make with α in order for its intersection with the cone to be a circle? A parabola? What angles produce hyperbolas? What angles produce ellipses?

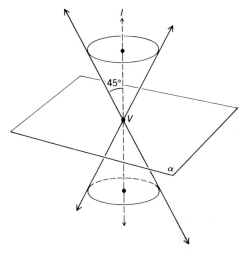

b) A degenerate point conic consists of a single point, one line, or two lines. Explain how planes can intersect the cone above to form these degenerate conics.

2. Explain how to sketch the various conics using definition (2) above and any type of materials you desire.

3. Sketch an ellipse, parabola, and hyperbola using definition (3) above.

4. Draw any straight line m. Locate a point F not on m. Fold the point F upon the line m. Repeat this fold 20 to 30 times by moving F along m. (Wax paper is very good for paper-folding activities.) What geometric figure is the set of lines represented by creases?

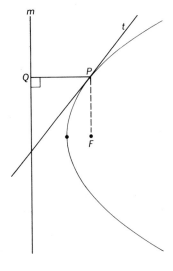

The paper-folding in the preceding exercise is based on the following theorem:

A tangent to a parabola bisects the angle between the line from the focus to the point of tangency and the perpendicular to the directrix from the point of tangency.

5. Draw a circle with center O. In the interior of the circle locate a point F different from O. Fold the point F upon the circle. Repeat this fold twenty to thirty times by folding F to different points on the circle, such as X and Y. What geometric figure is represented by the creases?

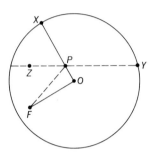

6. Explain why the paper-folding in the preceding example works. [*Hint:* Use the figure and recall that the crease $\overleftrightarrow{ZY}$ is the perpendicular bisector of $\overline{FX}$. Apply Definition (2) for an ellipse.]

7. Draw a circle with center O. Locate a point F in the exterior of the circle.

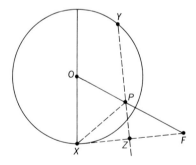

Fold F upon the circle repeatedly as above. What geometric figure is represented by the creases?

8. Explain why the paper-folding in the previous exercise works.

9. On graph paper, consider two pencils of lines on two distinct points, Q and Q'. Locate about twenty points which are intersections of perpendicular lines

(one in each pencil). What geometric figure is formed by the collection of all such points?

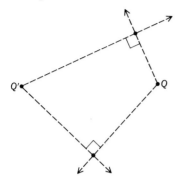

10. Consider the set of lines which are equidistant from a fixed point. Draw several of these lines. What point conic corresponds to this line conic?

11. Consider points Q and Q' and two nonperpendicular intersecting lines a and a' on Q and Q', respectively (see figure). What geometric figure is the set of points which are intersections of lines on Q and Q' forming congruent angles with lines a and a'? Locate several points before making a conjecture.

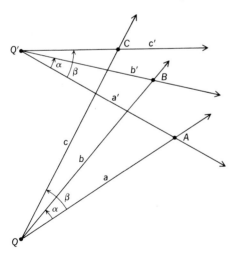

12. Consider the set of points which are intersections of lines on Q and Q' respectively, as in Exercise 11, but with the congruent angles made with line a in the opposite direction from those made with line a'. What geometric figure is formed by this set of points?

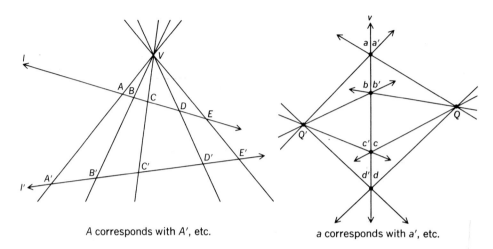

A corresponds with A′, etc.   a corresponds with a′, etc.

In the figure above, one-to-one correspondences between the points of l and l' and the lines on Q and Q' are shown. In projective geometry, these correspondences are called *projectivities* (projective transformations). A projectivity between the points of two lines is called a *perspectivity* or central projection if the lines determined by corresponding points are concurrent. A projectivity between the lines on two points is perspective if corresponding lines intersect in points which are collinear. The two projectivities above are perspective. We shall now show how to construct a nonperspective projectivity between the points of two distinct lines.

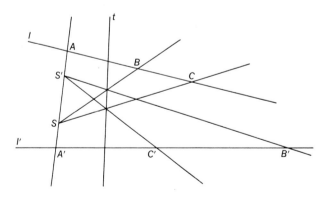

Choose a pair of points A, A' on l, l' respectively. Draw $\overleftrightarrow{AA'}$, and choose a pair of points S and S' on $\overleftrightarrow{AA'}$. Choose a line t not equal to l, l', or $\overleftrightarrow{AA'}$. Then B, B' (on l, l', respectively) correspond if $\overleftrightarrow{S'B'}$ and $\overleftrightarrow{SB}$ intersect on t. Clearly the correspondence is not perspective since $\overleftrightarrow{AA'}$, $\overleftrightarrow{BB'}$, and $\overleftrightarrow{CC'}$ are not concurrent. The correspondences of the lines on Q and Q' in Exercises 4 and 6 are examples of nonperspective correspondences between lines on two distinct points.

13. Find several more pairs of corresponding points in the nonperspective correspondence described above. Now use a different color and connect corresponding pairs of points. Construct a second nonperspective correspondence between the points of two lines, and draw the lines determined by pairs of corresponding points. What geometric figure is formed by this set of lines?

In Exercises 4 through 8, we see the foundation for another definition of point conics and line conics.

A *nondegenerate point conic* is the set of points which are intersections of corresponding lines in a nonperspective projectivity between two pencils of lines on distinct points.

A *nondegenerate line conic* is the set of lines determined by pairs of corresponding points in a nonperspective projectivity.

In both cases, we get a degenerate conic if the projectivity is perspective.

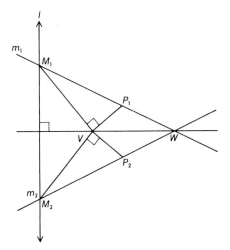

14. Consider two points V and W and a line $l \perp \overleftrightarrow{VW}$. Let m_1 be an arbitrary line on W meeting l at M_1. Draw $\overline{M_1V}$, and construct a perpendicular to $\overleftrightarrow{M_1V}$ at V. Let P_1 be the intersection of the perpendicular and m_1. In the figure above, a second point P_2 has been found. What geometric figure is the set of points determined in this manner?

15. Perform the construction of Exercise 14 with l meeting $\overline{VW}$. What happens if the distance from l to (V or W) is increased? What happens if the distance between V and W is increased?

16. Consider any conic and a triangle inscribed in the conic. Now consider the three points, each of which is the intersection of a tangent to the conic at a vertex of the triangle and the opposite side of the triangle (or the side extended). What appears to be true? Try this for other types of conics. Make a conjecture.

17. Consider an ellipse with an inscribed quadrilateral. Now consider the four points determined by the intersections of opposite sides of the quadrilateral and the intersections of tangents to the conic at opposite vertices. What appears to be true? Try this with a hyperbola, circle, and a parabola. Write a theorem which you think is true.

18. Draw an ellipse and a circumscribing hexagon. Draw the lines determined by opposite vertices of the hexagon.

   a) What appears to be true about these three lines?

   b) Try this for a circle and another ellipse.

   c) Make a conjecture.

The conjecture anticipated in Exercise 18 is known as *Brianchon's Theorem*. Compare this theorem with Pascal's Theorem.

10   Logic and Proof

10–1 Introduction

In the next chapter we shall indicate how geometry may be developed axiomatically. This will involve agreeing upon a few undefined terms and then listing some statements about these terms which we accept unquestioningly as "true." These statements will be our *axioms*. (The terms *assumption* and *postulate* have the same meaning as *axiom*.) Once the basic axioms are agreed upon, then the rules of logic take over. Of course our geometric intuition suggests the theorems that we attempt to prove, but our proofs must satisfy logical criteria. An argument is valid if it proceeds from basic assumptions to a conclusion in accordance with the laws of logic. Our chief purpose in studying logic in this chapter is to help the reader distinguish between valid and invalid arguments.

We give several examples below to illustrate the role of logic in proofs.

Example 1. A student in an algebra class was asked to prove the following theorem:

If a **and** b **are real numbers such that** $ab = 0$, **then** $a = 0$ **or** $b = 0$.

He said,

"I will restate the conclusion of the theorem. What I shall prove is that

"**If** $a \neq 0$,   **then** $b = 0$.

This is the same thing as proving that $a = 0$ or $b = 0$."

His classmates were not all convinced. Some said that he should also prove that if $b \neq 0$, then $a = 0$. What is your opinion?

Example 2. Consider the statement below:

It is not true that all men are honest.

Which statements below have exactly the same meaning as the preceding statement?

1. No men are honest.
2. All men are dishonest.
3. Some men are dishonest.
4. Some men are honest and some are dishonest.

Example 3. A student was asked to prove the theorem that there is no rational number whose square is 2. Before starting he explained his method for proof. He said,

"I will make the proof in the following three steps:

1. I will assume that there is a rational number r whose square is 2.
2. I know that every rational number can be represented by a lowest term fraction, and I will show that when this rational number r is written in lowest terms as a fraction a/b, then at least one of the two numbers a and b is odd.
3. I will prove that

$$\text{if} \quad \left(\frac{a}{b}\right)^2 = 2, \quad \text{then} \quad a \text{ and } b \text{ are both even.}$$

Then I will know that my assumption (1) is false, and the theorem will be proved." Do you consider this an acceptable plan of proof?

Example 4. A student was asked to prove the theorem:

If P is a point not on a line k, then there is at most one line perpendicular to k and containing P.

The student showed how to construct a line from P perpendicular to k. Had the student proved the assigned theorem?

These examples raise a basic question: How can we decide whether two statements that are worded differently have exactly the same meaning? We shall consider this question in the next section.

10–2 Logical Connectives

In Example 1 we considered two statements

1. **If** $ab = 0$,   **then** $a = 0$ **or** $b = 0$.
2. **If** $ab = 0$,   **then if** $a \neq 0$,   **then** $b = 0$.

Note the words, *if, then,* and *or.* These words connect simple statements to form complex statements. For example, in (1) we have three simple "atomic" statements,

$$ab = 0, \qquad a = 0, \qquad b = 0.$$

In (2) the statements are

$$ab = 0, \qquad a \neq 0, \qquad b = 0.$$

In everyday life we use many different connective words and phrases:

and, but, unless, because, . . .

In mathematics we use chiefly the three connectives,

and,   or,   if . . ., then . . .

If p and q are any two statements, we may form the statements:

p **and** q; p **or** q;   **if** p, **then** q.

For example, let p represent the false statement $3 + 4 = 8$, and let q be the false statement $(3 + 4) + 5 = 13$. Then

p **and** q is: $3 + 4 = 8$ **and** $(3 + 4) + 5 = 13$.

p **or** q is: $3 + 4 = 8$ **or** $(3 + 4) + 5 = 13$.

If p, **then** q is: **if** $3 + 4 = 8$, **then** $(3 + 4) + 5 = 13$.

Problem. In this example both atomic statements p and q are false. Complete the table below indicating by T (true) and F (false) your judgment of the truth value of each of the three compound statements:

p	q	p and q	p or q	if p, then q
F	F			

The mathematical usage of *and* agrees with everyday usage. A statement of the form

p *and* q

is true only when p is true and q is true.

The mathematical usage of *or* requires some explanation. In everyday life there are two distinct usages of *or*. A parent who says,

"You may have candy *or* ice cream,"

undoubtedly intends that his child will have one but *not* both.

A sentence of

"10 days in jail *or* a \$100 fine"

will not result in both penalties. But someone who comments,

"It will rain today or tomorrow,"

probably intends to assert that it will rain on one and possibly on both of these days.

In legal terminology, ambiguity is avoided by using two connectives, *or* and *and/or*. If the law prescribes a punishment of

"10 days in jail and/or a \$100 fine,"

then the judge has three different penalties that he may inflict. (What are they?)

In mathematics, ambiguity is avoided by assigning to *or* the meaning of the legal *and/or*. That is,

p or q

is considered to be false only when p is false and q is false. The *mathematician* who promises his child candy or ice cream is not violating his promise if he accedes to pressures and gives him both.

The usages of *and* and *or* are neatly summarized by the truth table below. This table indicates that for any statements p, q: the truth and falsity of the complex statements, p and q, p or q, are as shown.

p	q	p and q	p or q
T	T	T	T
T	F	F	T
F	T	F	T
F	F	F	F

Statements of the form:

if p,   then q,

are called *implications* and may be phrased as

p implies q.

Often a hollow arrow replaces the word "implies" and one writes

$p \Rightarrow q.$

It is obvious that statements of implication play an important role in mathematical proof. For example, we may think of the task of solving an equation as a matter of setting up an appropriate string of implications.

$$2X + 3 = 13 \Rightarrow 2X = 10 \Rightarrow X = 5.$$

In the *if . . . then* language, this argument runs:

If $2X + 3 = 13$,   *then* $2X = 10$.

If $2X = 10$,   *then* $X = 5$.

Note that from these two implications we deduce a third:

If $2X + 3 = 13$,   *then* $X = 5$.

The example above illustrates a rule of logic which has the following symbolic expression:

$$[(p \Rightarrow q) \text{ and } (q \Rightarrow r)] \Rightarrow (p \Rightarrow r).$$

This is one of the chief ways in which implication is used in everyday reasoning. We see that some first proposition implies a second, the second implies a third, and so we recognize that the first implies the third.

The example below illustrates another use of implication in constructing proofs. As the figure indicates,

$$\overline{AB} \cong \overline{AC} \qquad \text{and} \qquad \overline{DB} \cong \overline{DC}.$$

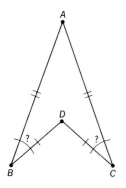

Our task is to prove that

$$\angle B = \angle C.$$

We reason as follows:

If $\triangle ABD$ is congruent to $\triangle ACD$, then $\angle B \cong \angle C$. $\triangle ABD$ is congruent to $\triangle ACD$ by the SSS theorem. Hence, $\angle B \cong \angle C$. Symbolically:

$p \Rightarrow q$

$\dfrac{p}{q}$

This rule of logic is known as the *law of detachment*. If an implication, $p \Rightarrow q$ is true, and if the statement p is true, then also q is true.

Another use of implication closely related to the law of detachment is described below. Suppose that we are attempting to prove that

$$\angle 1 \not\cong \angle 2.$$

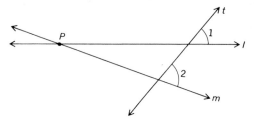

We reason as follows:

  1. $\angle 1 \cong \angle 2 \Rightarrow l \parallel m.$ (Why?)   2. $l \not\parallel m.$ (Why?)
  3. Hence, $\angle 1 \not\cong \angle 2.$

Symbolically we have

$$p \Rightarrow q$$
$$\mathrm{not}\ q$$
$$\overline{\phantom{\mathrm{not}\ q}}$$
$$\mathrm{not}\ p$$

If an implication, $p \Rightarrow q$, is true, and if q is false, then p cannot be true. This sort of proof is often called an *indirect argument*, and we say that we *assume* the truth of p, arrive at a statement known to be *false*, and then know that p is false.

We give the truth table for implication below.

p	q	$p \Rightarrow q$
T	T	T
T	F	F
F	T	T
F	F	T

This table is best understood by focusing attention upon the second line. Note that the table can be described briefly as

$p \Rightarrow q$ is false when p is true and q is false. In all other cases $p \Rightarrow q$ is true.

It is relatively easy to justify this property of the table. For example, suppose that one of two tennis players asserts,

If I lose this set of tennis, *then* I'll eat my hat.

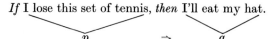

$$p \qquad\qquad \Rightarrow \qquad\qquad q$$

Suppose also that you learn sometime later that his statement of implication is *false*. Then it is clear that

1) He lost the set.   2) He failed to eat his hat.

Knowing that $p \Rightarrow q$ was false, you deduced that p was *true* and q, *false*. Hence this must be the only condition under which $p \Rightarrow q$ is a false statement.

You will remember that the table for *or* was also true in three of the four possible cases and false in one. This suggests a possible connection between statements of implication and *or*-statements. If we denote by "*not p*" the statement "*p is false*," then we have the following truth table.

p	q	$p \Rightarrow q$	*not p*	(*not p*) **or** q
T	T	T	F	T
T	F	F	F	F
F	T	T	T	T
F	F	T	T	T

Explain how the table above shows that the statements

$$p \Rightarrow q; \qquad (\text{not } p) \text{ or } q$$

have exactly the same meaning.

We can settle the question raised in Example 1 at the beginning of this chapter. In order to decide whether the two statements,

1. *If* $ab = 0$,   *then* $a = 0$ *or* $b = 0$,
2. *If* $ab = 0$,   *then if* $a \neq 0$,   *then* $b = 0$,

have the same meaning, we need only examine the statements

i) $a = 0$ or $b = 0$ ii) If $a \neq 0$,   then $b = 0$.

Denoting the statement $a \neq 0$ by p, and $b = 0$ by q, we may rewrite these as

i') (not p) or q,   ii') $p \Rightarrow q$.

The truth table above shows that these statements have the same meaning, and so the approach to the proof taken by the student was correct.

We have used the rather vague phrase, *have the same meaning*, in comparing different statements. More precisely, we say that two statements are equivalent if each statement implies the other.

p is *equivalent* to q means: $p \Rightarrow q$ and $q \Rightarrow p$.

We use a double arrow to indicate this relationship of equivalence between two statements:

$$p \Leftrightarrow q.$$

We have remarked that one reason we study logic systematically is to increase our ability to distinguish between valid and invalid arguments. But a knowledge of logic also provides guidance in developing a plan of proof. Often before solving a problem we must cast about for possible methods of attack. The selection of a particular method may be strongly suggested by an understanding of logical principles.

Exercises 10–1

1. Copy and complete the truth table below.

p	q	p and q	p or q	$p \Rightarrow q$	$p \Leftrightarrow q$
T	T				
T	F				
F	T				
F	F				

2. Verify from the truth table of Exercise 1 that the statement $p \Leftrightarrow q$ is true

   a) when both statements are true,

   b) when both statements are false;

   and false otherwise. Hence, equivalent statements can be thought of as statements which are either both true or both false.

3. A student remarked, "We will win the baseball game Friday or the track meet Saturday." It develops that this student made a false statement. What actually happened?

4. Repeat Exercise 3 replacing the *or* of the student's statement by *and*.

5. Verify that the two statements, p or q, q or p are equivalent.

6. What can you deduce if you know

   a) not (p or q) [i.e., you know the statement, p or q is false]

   b) not (p and q)   c) not ($p \Rightarrow q$)

   d) not ($q \Rightarrow p$)   e) not ($p \Leftrightarrow q$)

7. A man promised a friend: "*If* I come to town Thursday, *then* I shall telephone you." Judge this statement as true or false under the following conditions:

   a) He comes to town and telephones his friend.

   b) He comes to town and does not telephone his friend.

   c) He does not come to town and telephones his friend.

   d) He does not come to town and does not telephone his friend.

8. It is fairly obvious that when a stern parent remarks, "You will wash the dishes tonight *or* you will not go to the party tomorrow night," he means, "If you do not wash the dishes tonight, then you will not go to the party tomorrow night." Explain how this illustrates the logical principle, (not p) or $q \Leftrightarrow$ (if p, then q).

9. Construct an *if ... then ...* statement equivalent to each *or* statement below.

   a) You will sit quietly or I will spank you.

   b) $x = 2$ or $x = 4$.

   c) $x = 4$ or $x = 2$.

   d) I will spank you or you will sit quietly.

   e) p or q f) p or p

   g) (not p) or q h) (Not p) or (not q)

10. Construct *or* statements equivalent to the *if ... then ...* statements below.

   a) If he does not telephone soon, then I shall be angry.

   b) If it turns cold, then it will rain.

   c) If $x \neq 4$, then $x = 5$.   d) If $x = 4$, then $x = 5$.

   e) If $x = 5$, then $x \neq 4$.   f) If $l \nparallel m$, then $l \nparallel n$.

   g) If $l \parallel m$, then $l \parallel n$.   h) If p, then q.

   i) If q, then p.   j) If (not p), then q.

   k) If q, then (not p).   l) If (not p), then (not q).

   m) If (not q), then (not p).   n) If (not q), then (not q).

   o) If q, then (not q).

Exercise 9 above calls attention to the influence of language habits upon our understanding of logical concepts. We are accustomed to phrasing statements in certain stereotyped ways. It is quite easy to recognize that the statement of 9(a),

   You will sit quietly *or* I will spank you,

which has the form

   p or q

can be stated in equivalent fashion as

   If you will *not* sit quietly, *then* I will spank you.

This has the form

  if (not p), *then* q.

But any statement of the form *p or q* is equivalent to *q or p*, and so a second equivalent form of this statement is

  if (not q), *then* p.

If I do not spank you, *then* you will have sat quietly.

Because this last statement is an unusual way of speaking, it is difficult to recognize the equivalence of all these forms. Observe that we have four different forms of statements, all equivalent to *p or q*.

  1. *p or q*,   2. if not *p*, then *q*,   3. *q or p*,   4. if not *q*, then *p*.

Problem. Explain the ways in which you think of the four expressions above and recognize their equivalence.

It is advantageous, when studying logic, to focus some attention upon formal relationships that have nothing to do with the *content* of individual statements. One way to do this is by proving some *theorems* using statement variables, p, q, ... rather than particular statements. We illustrate this technique in examples below. For each theorem we list a *hypothesis*, a set of statement expressions each of which we consider to be "true," and then we use our knowledge of logical laws to deduce other "true" statement expressions. Explain for each example how the conclusion is deduced.

Example 1
  Hypothesis: *p or q*; not *q*
  Conclusion: *p*

Example 2
  Hypothesis: $p \Rightarrow$ not q; p
  Conclusion: not q

Example 3
  Hypothesis: $p \Rightarrow$ (*q or r*); $r \Rightarrow$ not p, $q \Rightarrow r$
  Conclusion: not p

Exercises 10–2

  1. Explain your proofs of all theorems below.

    a) Hypothesis: $p \Rightarrow q$; not q
      Conclusion: not p

b) Hypothesis: $p \Rightarrow$ not q; q
   Conclusion: not p

c) Hypothesis: not $q \Rightarrow$ not p; not q
   Conclusion: not p

d) Hypothesis: $p \Rightarrow q$; $q \Rightarrow r$; p
   Conclusion: q and r

e) Hypothesis: $p \Rightarrow q$; $q \Rightarrow r$; $r \Rightarrow$ not p
   Conclusion: not p

f) Hypothesis: (not p) or q; (not q) or r; (not r) or (not p)
   Conclusion: not p

g) Hypothesis: $p \Rightarrow (q \Rightarrow r)$; p; not r
   Conclusion: not q; not $(p \Rightarrow q)$; not $[p \Rightarrow (q$ or $r)]$; $r \Rightarrow q$; $p \Rightarrow (r \Rightarrow q)$

h) Hypothesis: $p \Rightarrow q$
   Conclusion: not $q \Rightarrow$ not p

i) Hypothesis: not (p or q)
   Conclusion: not p; not q; $p \Leftrightarrow q$

j) Hypothesis: $p \Rightarrow (q$ or $r)$
   Conclusion: $(p \Rightarrow q)$ or $(p \Rightarrow r)$

k) Hypothesis: $p \Rightarrow q$; $q \Rightarrow r$; $r \Rightarrow p$
   Conclusion: $p \Leftrightarrow q$; $q \Leftrightarrow r$; $r \Leftrightarrow p$

2. Explain why each of the following suggested conclusions is not derivable from the given hypothesis.

a) Hypothesis: $p \Rightarrow q$; q
   Conclusion: p

b) Hypothesis: $p \Rightarrow q$
   Conclusion: $q \Rightarrow p$

c) Hypothesis: $p \Rightarrow (q \Rightarrow r)$; p; r
   Conclusion: q

3. Show how to set up truth tables and solve the problems of Exercises 1 and 2 mechanically.

4. Draw conclusions from each of the following hypotheses:

a) not (p and q); q

b) not (p and not q); p

c) $p \Rightarrow q$; $q \Rightarrow$ not p

d) $p \Rightarrow q$; r or not q; not r

5. Let q be the statement, "The sum of the angles of every triangle is a straight angle." Let p be the statement, "There is a line l and a point P not on l such that there are two distinct perpendicular lines m and n from P to l." Assume that q is true. Show that the statement $p \Rightarrow$ not q is true, and draw a conclusion.

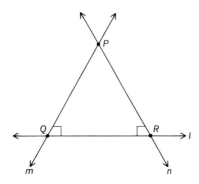

6. Let $p, q, r, \ldots$ be the following statements.

p: The sum of the measures of the angles of every triangle is $179°$.

q: The sum of the measures of the angles of $\triangle ABD$ is the sum of the measures of $\angle 1$, $\angle 2$, $\angle 3$, and $\angle 4$.

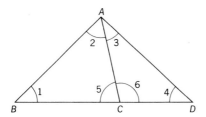

r: The sum of the measures of $\angle 1$, $\angle 2$, $\angle 3$, $\angle 4$, $\angle 5$, and $\angle 6$ is $358°$.

s: The sum of the measures of $\angle 5$ and $\angle 6$ is $180°$.

Assume that q and s are true. Show that the following statements are true:

$p \Rightarrow r$;   (p and q and s) $\Rightarrow$ not r.

What can you conclude?

10–3 Negation and Quantifiers

If p is any statement, the statement, not p, is called the *negation* of p. In the exercises of the preceding section you have had opportunities to observe several properties of negation. Explain the equivalences below.

  a) not (not p) $\Leftrightarrow p$

  b) not (p and q) $\Leftrightarrow$ (not p) or (not q)

  c) not (p or q) $\Leftrightarrow$ (not p) and (not q)

  d) not ($p \Rightarrow q$) $\Leftrightarrow p$ and (not q)

  e) not ($p \Leftrightarrow q$) $\Leftrightarrow$ (p and not q) or (q and (not p))

  f) not ($p \Leftrightarrow q$) $\Leftrightarrow$ (not p or (not q)) and (p or q)

The negation of simple statements like

$$x \in A; \; x = 3; \; x < 7; \; l \parallel m; \; \overline{AB} \cong \overline{CD}; \; A \subset B; \ldots$$

are conventionally indicated by drawing a slash mark through the symbol of relationship

$$x \notin A; \; x \neq 3; \; x \not< 7; \; 1 \not\parallel m; \ldots$$

Often, however, we need to negate statements of the following form:

  a) *For all* $x \in A$, x is even.

  b) *There exists* an $x \in A$, such that x is even.

Note that (a) is the assertion that *every* element of set A has a certain property. To negate this statement is to assert that there *is at least one* element of A which does *not* have the property. Statement (b) asserts that *at least one* element of A has a certain property. To negate (b) we must say that *no* element of A has the property. Alternatively, we may say that *every* element fails to have the property. Hence, negations of (a) and (b) may be expressed as

  a') *There exists* an $x \in A$ such that x is *not* even.

  b') *For all* $x \in A$, x is *not* even.

The phrases *for all* and *there exists* are called quantifiers. Standard abbreviations for these quantifiers are

  $\forall$ (for all); $\exists$ (there exists).

The symbol $\forall$ is called the *universal quantifier*, and $\exists$ is called the *existential quantifier*. Thus statements (a), (a'), (b), (b') may be written as

  a) $\forall x \in A$, x is even;   a') $\exists x \in A$ such that x is not even;

  b) $\exists x \in A$ such that x is even;   b') $\forall x \in A$, x is not even.

Note that the negation of a statement involving the universal quantifier is one involving the existential quantifier, and vice versa. Many of the important statements of geometry employ these quantifiers although the language may conceal their use. For example, the fundamental axiom,

> *Two points determine a straight line*, when spelled out precisely is: "*For all* points A, B, if $A \neq B$, then *there exists* one and only one line l such that $A \in l$ and $B \in l$."

In algebra careful statements of the associative, commutative, distributive, etc., laws require use of the universal quantifier. For example, in the set W of whole numbers we have:

$\forall a, b \in W, a + b = b + a.$

$\forall a, b, c \in W, a(b + c) = ab + ac.$

$\forall a, b, c \in W, (a + b) + c = a + (b + c).$

In order to describe the properties of the identity elements 1 and 0 we need to use both quantifiers.

$\exists 0 \in W$ such that $\forall a \in W, a + 0 = a.$

$\exists 1 \in W$ such that $\forall a \in W, a \cdot 1 = a.$

Exercises 10–3

1. Express each statement below in a form in which the word *not* is not the first word of the sentence.

   a) Not (Roses are red and violets are blue.)

   b) Not (Roses are red or violets are blue.)

   c) Not (If roses are red, then violets are blue.)

2. Negate each statement below.

   a) All cats are black.

   b) Every Chinese is short.

   c) Every Frenchman is honest.

   d) $\forall x \in U$, x is a squirrel.

   e) $\forall x \in U$, x is good and brave.

   f) $\forall x \in U$, if $x > 20$, then $x < 30$.

   g) There exists a tall Chinese.

   h) There is a cat in the pound which has no tail.

   i) $\exists x \in U$ such that x stammers.

   j) $\exists x \in U$ such that x is even and $x > 40$.

k) $\exists x \in U$ such that $\forall y \in U, y \neq x \Rightarrow y < x$.

3. Let U be the set $\{1, 2, 3, 4, 5, 6, 7, 8, 9, 10\}$. Decide whether each statement below is true or false.

a) $\forall x \in U, x > 9 \Rightarrow x$ is even.

b) $\forall x \in U, x < 3 \Rightarrow x$ is odd.

c) $\exists x \in U$ such that $x + 5 = 9$.

d) $\exists x \in U$ such that $x - 8 = 4$.

e) $\exists x, y \in U$ such that $x + y = 11$ and $x - y = 3$.

f) $\exists x \in U$ such that $\forall y \in U$, if $y \neq x$, then $y > x$.

g) $\forall x \in U, \exists y \in U$ such that $y > x$.

4. Observe that Exercise 1 is ambiguous because it is not clear whether "Roses are red" means

a) $\forall x$, if x is a rose, then x is red or

b) $\exists x$ such that x is a rose and x is red.

Rework parts (a) and (b) of Exercise 1 using both these interpretations for roses and violets.

5. Let L be the set of all lines of the Euclidean plane E. Decide whether each statement below is true or false. We use the usual notation, small letters for lines and capitals for points.

a) $\forall l, m \in L, l \not\parallel m \Rightarrow \exists P \in E$ such that $P \in l$ and $P \in m$.

b) $\forall l, m \in L, l \perp m \Rightarrow m \perp l$.

c) $\forall l, m, n \in L, l \perp m$ and $m \parallel n \Rightarrow l \perp n$.

d) $\forall l \in L, \forall P \in E, \exists m \in L$ such that $P \in m$ and $m \perp l$.

e) $\forall l, m, n \in L, l \not\parallel m$ and $m \not\parallel n \Rightarrow l \not\parallel n$.

f) $\forall l \in L, \forall P \in E, \exists m \in L$ such that $P \in m$ and $m \parallel l$.

g) $\forall l, m, n, p \in L, l \perp m$ and $m \perp n$ and $n \perp p \Rightarrow p \perp l$.

h) $\forall A, B, C, D \in E, \overleftrightarrow{AB} \parallel \overleftrightarrow{CD}$ and $\overline{AC} \cong \overline{BD} \Rightarrow \overleftrightarrow{AC} \parallel \overleftrightarrow{BD}$ and $\overline{AB} \cong \overline{CD}$.

i) $\forall A, B, C, D \in E, \overleftrightarrow{AB} \parallel \overleftrightarrow{CD}$ and $\overleftrightarrow{AC} \parallel \overleftrightarrow{BD} \Rightarrow \overline{AB} \cong \overline{CD}$ and $\overline{AD} \cong \overline{BC}$.

j) $\forall A, B \in E$ such that $A \neq B, \exists P \in E$ such that $P \in \overleftrightarrow{AB}$ and $\overline{AP} \cong \overline{BP}$.

k) $\forall A, B, C, D \in E, \overline{AC} \cong \overline{BC}$ and $\overline{AD} \cong \overline{BD}$ and $A \neq B$ and $C \neq D \Rightarrow \overleftrightarrow{AB} \perp \overleftrightarrow{CD}$.

l) $\forall A, B, C \in E, \overline{AB} \cong \overline{AC} \Rightarrow \angle ABC \cong \angle ACB$.

10–4 Converses and Contrapositives of Implications

With each statement of implication, $p \Rightarrow q$, it is natural to associate two other statements, one called the converse of $p \Rightarrow q$ and the other the contrapositive of $p \Rightarrow q$. The converse of $p \Rightarrow q$ is $q \Rightarrow p$. The contrapositive of $p \Rightarrow q$ is not $q \Rightarrow$ not p. These concepts are important because of the roles they play in proofs. If $p \Rightarrow q$ and $q \Rightarrow p$ (its converse) are both true, then p and q are equivalent statements and may be employed interchangeably. For example, in any triangle $\triangle ABC$,

a) $\overline{AB} \cong \overline{AC} \Rightarrow \angle B \cong \angle C$, and also

b) $\angle B \cong \angle C \Rightarrow \overline{AB} \cong \overline{AC}$.

Each of (a) and (b) is the converse of the other. We may combine (a) and (b) in a single statement of equivalence

c) $\overline{AB} \cong \overline{AC} \Leftrightarrow \angle B \cong \angle C$.

Statements of equivalence are often phrased as *if and only if* statements.

c′) A triangle is isosceles *if and only if* its base angles are congruent.

Or we may say

c″) If a triangle is isosceles, then its base angles are congruent and conversely.

Note that

p if q means: if q then p, i.e., $q \Rightarrow p$.

p only if q means if p then q, i.e., $p \Rightarrow q$.

A natural interpretation of a statement, *p only if q*, is

if not q *then* not p.

This last statement is the *contrapositive* of the implication $p \Rightarrow q$, and as the development above indicates the contrapositive of an implication is equivalent to the implication itself.

$(p \Rightarrow q) \Leftrightarrow (\text{not } q \Rightarrow \text{not } p)$.

One of the major uses of this relationship between an implication and its contrapositive is that when faced with a theorem of the form

$p \Rightarrow q$,

we may replace it by the contrapositive form, not $q \Rightarrow$ not p. As an example, consider the theorem

$\forall a \in W$,   if a^2 is even, then a is even.

It is more straightforward to prove the contrapositive.

$\forall a \in W$,   if a is odd, then a^2 is odd.

Exercises 10–4

1. Show that each statement below is true. In each statement replace the implication by its converse and show that the resulting statement is false.

   a) For all quadrilaterals $ABCD$, if $ABCD$ is a square, then $AC \perp BD$.

   b) For all triangles ABC, if $\triangle ABC$ is equilateral, then $\overline{AB} \cong \overline{BC}$.

   c) For all whole numbers a, b, if a and b are both even, then $a + b$ is even.

   d) For all whole numbers x, y, $x = 3$ and $y \to 5 \Rightarrow x + 2y = 13$.

   e) For all real numbers x, $x > 0 \Rightarrow x^2 > 0$.

2. For each part of Exercise 1, replace the implication by its contrapositive and satisfy yourself that the resulting statement is true.

3. Show that the converse of each implication below is true. Which of the implications are true?

   a) For all quadrilaterals, if a quadrilateral is a parallelogram, then its diagonals bisect each other.

   b) For all quadrilaterals, if a diagonal bisects each of two opposite angles, then the quadrilateral is a square.

   c) For all quadrilaterals, if the diagonals are congruent, then the quadrilateral is a rectangle.

   d) For all whole numbers x, y, $x = 4$ and $y = 3 \Rightarrow x + y = 7$ and $3x + y = 15$.

4. Show that each implication below is true, state its converse, and verify that the converse is true also.

   a) $\forall a, b \in W$, $a > b \Rightarrow a + 2 > b + 2$.

   b) $\forall m, n \in I$, $m^2 = n^2 \Rightarrow m = n$ or $m = -n$.

   c) For all triangles ABC, if $\overline{AB} > \overline{AC}$, then $\angle C > \angle B$.

   d) For all triangles ABC and DEF, if $\triangle ABC \cong \triangle DEF$, then $\overline{AB} \cong \overline{DE}$, $\angle B \cong \angle E$, and $\overline{BC} \cong \overline{EF}$.

   e) For all real numbers x, y, $2x + 3y = 7$ and $5x - 4y = 11 \Rightarrow x = \frac{61}{23}$, and $y = \frac{13}{23}$.

5. State the contrapositive of each theorem.

   a) $\forall a, b \in W$, $ab = 0 \Rightarrow a = 0$ or $b = 0$.

   b) For all quadrilaterals, if the diagonals are not perpendicular, then the quadrilateral is not a rhombus.

   c) If a transversal of two lines forms a pair of congruent corresponding angles, then the lines are parallel.

   d) If a triangle is isosceles, then it has at least one pair of congruent angles.

   e) If a triangle is equilateral, then its circumcenter and incenter coincide.

f) For all convex quadrilaterals $ABCD$, if $\angle A$ is not supplementary to $\angle C$, then no circle circumscribes $ABCD$.

Sometimes the terms *necessary* and *sufficient* are used in referring to implications. The accepted usage of the word sufficient is intuitively clear. To say for two statements p and q that

p is a *sufficient* condition for q means simply that

$p \Rightarrow q$.

To say that p is a *necessary* condition for q is to assert

$q \Rightarrow p$.

It is easy to confuse the meaning of these terms. It may help you remember the meaning of necessary if you interpret

p is necessary for q as   not $p \Rightarrow$ not q.

Replacing this implication by its contrapositive we have,

$q \Rightarrow p$.

"If and only if" theorems are often stated as "necessary and sufficient" theorems. For example, a triangle is isosceles

if and only if two of its angles are congruent.

A *necessary and sufficient* condition that a triangle be isosceles is that two of its angles be congruent.

Exercises 10–5

1. Does the statement $p \Rightarrow q$ mean p *only if q*, or does it mean q *only if p*?

2. Does the statement $p \Rightarrow q$ mean p is *necessary* for q, or does it mean q is *necessary* for p?

3. Does the statement $p \Rightarrow q$ mean p is *sufficient* for q, or does it mean q is *sufficient* for p?

4. Consider the theorem:

   A quadrilateral is a parallelogram if and only if its diagonals bisect each other.

   a) In proving the "if" part of this theorem, what statement is taken as the hypothesis?

   b) In proving the "only if" part of this theorem, what statement is taken as the hypothesis?

5. Consider the theorem:

In order for a quadrilateral to be a parallelogram, it is necessary and sufficient that its diagonals bisect each other.

a) In proving necessity, what is the hypothesis?

b) In proving sufficiency, what is the hypothesis?

6. The Venn diagram indicates that the set of all brave men B is a subset of all honest men H. Which of the statements below agree with this Venn diagram?

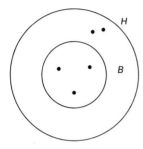

a) If a man is brave, then he is honest.

b) If a man is honest, then he is brave.

c) If a man is not brave, then he is not honest.

d) If a man is not honest, then he is not brave.

e) A man is brave only if he is honest.

f) A man is honest only if he is brave.

g) In order to be brave it is necessary that a man be honest.

h) In order to be honest it is necessary that a man be brave.

i) Bravery is a sufficient condition for honesty.

j) Honesty is a sufficient condition for bravery.

k) A man is not brave only if he is not honest.

l) A man is not honest only if he is not brave.

11　The Axiomatic Structure
of Geometry

11–1　Introduction

So far, our treatment of geometry has been intuitive. We have made many "proofs," but we have not formulated a complete set of axioms upon which these proofs can be based. Euclid was the first mathematician to organize a fairly satisfactory list of assumptions for the development of geometry. It is a tribute to his genius that his work stood almost intact for 2000 years. But in the late 19th century, mathematicians carefully reviewed the geometry of Euclid and pinpointed its shortcomings. By 1900 it was known how to place geometry upon a firm logical basis. That is to say, mathematicians were able to list axioms from which all the thousands of known geometric theorems (and the millions as yet undiscovered!) could be proved. With these axioms one does not need to point at diagrams. All proofs can be made verbally according to the rules of logic. From this point of view the theorems of geometry arise in the following way:

> The axioms are basic statements $p, q, r, \ldots$ which we consider to be true. The laws of logic furnish us with techniques for constructing new true statements from these basic ones. These new statements are our theorems.

Of course, in actual practice most geometric theorems are discovered intuitively. One visualizes geometric figures, guesses that a certain statement is true, and then turns to the axiomatic structure in an attempt to verify the guess.

Our purpose in this chapter is not to develop geometry in the precise rigorous way described above, but rather to give the reader some of the flavor of an axiomatic development. For the most part the theorems considered will be quite simple. Several of the logical concepts introduced in Chapter 10 will be helpful, but our use of logic will be naive and intuitive.

11–2 Undefined Terms

The inevitability of undefined terms is clear. In any discourse it is impossible to define every word in terms of other words. Any attempt to do so inevitably results in encountering a circular chain of definitions. If you choose any word in a dictionary and pursue its synonyms tirelessly through the pages, eventually you come to a word without a given synonym, or you travel in a full circle. Since we have only a finite number of words in our language, it cannot be otherwise. In geometry the terms,

> *point, line, plane, space*

are usually taken as undefined. The axioms establish *relationships* between these undefined entities. For example, we may assume:

> *Space is the set of all points.*

This axiom does not tell us what space is or what points are. Rather it describes a relationship between points and space. Now we know that **if** P is any **point** of our geometry, **then** P is an element of the set called space.

Some axioms are *existence* axioms. These assure us that we really have a basis on which to prove theorems. As an example of an existence axiom, we might assume:

> *Space contains at least four points.*

Exercises 11–1

1. For each of the words "house" and "honesty," track synonyms through a dictionary, forming as long a circular chain as possible.

2. In real life we could convey to someone who had never seen a horse an understanding of what one is by (a) showing him horses, (b) showing him pictures of horses, or (c) explaining to him what horses do, that is, describing how they interact with their environment. Discuss the possibility of describing points and lines in this way.

11–3 A Finite Geometry

In order to emphasize the role of axioms in the development of a mathematical system, we present in this section a collection of axioms that determines a *finite* geometry, that is, a geometry that contains only finitely many "points." This will be a "plane" geometry. In other words, all our points will be postulated to lie in one set called a *plane*. Our undefined terms will be *point, line,* and *plane.* All proofs of theorems will be left to the reader. If you work through the suggested problems and theorems carefully, the following statements should become a bit more meaningful.

a) The basic elements of any mathematical system are undefined.

b) The basic axioms of any mathematical system are statements about the undefined elements which are to be accepted unquestioningly as "true." We may formulate any axioms we please so long as they do not lead to contradictory conclusions.

c) The theorems of any mathematical system are derived from our axioms by using the laws of logic to form new "true" statements from these axioms.

In the sequential development below try to work all problems and prove all theorems. Draw diagrams freely or use any other notational devices to assist your reasoning. Note that from our axioms it will develop that each "line" will be a *finite* set of points. So, if you draw the ordinary ink streak pictures for these lines, don't make the mistake of assuming that if two ink streaks cross, the intersection of the marks must represent a point.

Undefined Terms: point, line, plane.

Axiom 1. We assume that there is a set called a *plane*. All elements of this plane are called *points*. Each *line* is a subset of this plane. There are at least 3 and no more than 7 points in the plane.

Problem 1. Using Axiom 1, can you prove that there are no more than 8 points in the plane? No more than 6 points? At least 4 points?

Axiom 2. For any two distinct points in the plane there is a unique line which contains them.

Problem 2. Can you prove the existence of at least one line? At least two lines?

Problem 3. Do the two axioms we have stated preclude the possibility that some line might consist of a single point? That some line might be the empty set? Draw "pictures" of at least three different geometries that satisfy our axioms. For example, let one of your figures show all the points of our geometry on a single line. Draw one diagram showing that there could be exactly two lines in our plane.

Axiom 3. No line contains all the points of the plane.

Theorem 1. There are at least 3 lines.

Instead of saying "a point is an *element* of a line," we will say "a point is on a line." Similarly, "a line is on a point" will mean that the line *contains* the point.

Axiom 4. Every line is on at least 3 points.

Definition 1. Two lines are *parallel* if and only if their intersection is the empty set.

Theorem 2. No two lines are parallel.

Theorem 3. Every pair of distinct lines intersects in exactly one point.

Theorem 4. Every line is on exactly 3 points.

Theorem 5. No point is on 4 lines.

Theorem 6. Every point is on at least 3 lines.

Theorem 7. There are 3 points on every line and 3 lines on every point.

Theorem 8. There are exactly 7 points.

Theorem 9. There are exactly 7 lines.

Definition 2. A triangle is a set of 3 noncollinear points.

Problem 4. How many triangles are there?

Problem 5. State and prove some theorem of your own concerning this geometry.

Problem 6. Is it possible to picture this geometry on a sheet of paper, picturing points as dots, so that the 3 points on every line lie on a straight mark on the paper?

Now read again the statements (a), (b), and (c) at the beginning of this section.

11–4 Axiom Systems for Plane Geometry

In Chapter 2 we referred to Euclid's axioms for geometry. We have mentioned that these are not logically satisfactory, but this blemish does little to mar the intrinsic beauty of Euclid's approach. Euclid's axioms agree with our spatial intuition. They assert things about physical space that we are willing to accept on a common-sense basis. Modern refinements of Euclid correct his blunders in logic, but these mathematically sound presentations necessarily lose some of the simplicity that makes Euclid attractive. For example, Euclid has an error in the proof of Proposition 1, Book I, in which he describes the construction of an equilateral triangle on a given segment. Euclid says: "Let the circle with A as center and radius AB be drawn and also the circle with B as center and radius *BA. Let C be a point in which these circles intersect ....*"

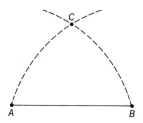

We have printed the sentence above in italics since this pinpoints one of Euclid's basic omissions. It is not possible to prove, using only Euclid's axioms that these circles really do have a point in common. In order to know that there is a point on both circles, we need to know that there *is* a point in the plane equidistant from A and B and also just as far from A as B is. Euclid's assumptions do not suffice to prove this. Of course, the *marks* we *draw* to represent these circles cross, but this is no proof that the circles intersect. Proofs in geometry are supposed to be based upon *verbal statements*, not upon pictures.

As another example of a logical omission, Euclid uses properties of *betweenness* without laying a justifying axiomatic foundation. Ideas of *greater than* and *less than* depend upon concepts of *betweenness*. Referring to the figure below, Euclid would say that $AB > AC$ because AC is **part** of AB and the whole is greater than any of its parts.

If we ask what it means for AC to be a "part" of AB, we will be told that C is **between** A and B. But if we turn to Euclid's axioms we find that no properties of this relation designated by the word "between" are presented by the axioms.

Although mathematicians have known for seventy years how to correct Euclid's errors, there has been, until recently, little effort to base courses in high school geometry upon a logically sound set of postulates. We list three such postulate sets in this section. It may be that in your high-school geometry course, you studied from some modern text that used one of these axiomatic systems, but it is possible that the text you used employed a collection of axioms with more deficiencies in logic than those of Euclid.

Before Euclid lists his "postulates" and "common notions" he makes several definitions. Several of these are futile. For example,

1. A *point* is that which has no *part*.

2. A *line* is *breadthless length*.

We have seen that some basic elements of geometry must be left undefined, and Euclid would have shown better judgment if he had omitted these definitions. "Breadthless length" seems a more complex concept than "line." Contrast with these definitions the introductory remarks below of David Hilbert as he introduces for the first time (1899) a satisfactory axiom system for geometry.

"Let us consider three distinct systems of things. The things composing the first system we shall call *points*, and designate them by the letters $A, B, C, \ldots$, those of the second we will call *straight lines* and designate them by letters $a, b, c, \ldots$, and those of the third system we will call *planes* and designate them by the Greek letters $\alpha, \beta, \gamma, \ldots$. The points are called the *elements of linear geometry*; the points and straight lines, the *elements of plane geometry*; and the points, lines, and planes, the *elements of space*."

Before listing Hilbert's postulates, we present Euclid's. For the postulates of Hilbert are best understood when one considers them as filling the logical omissions of Euclid.

Euclid's Postulates

1. A line (segment) can be drawn from any point to any point.
2. A line (segment) can be extended any desired distance.
3. A circle can be drawn with any center and radius.
4. Any two right angles are equal.
5. If a line intersects two lines and forms interior angles on one side with a sum less than two right angles, then the two lines, if extended, will meet on that side.
6. If equals are added to equals, the sums are equal.
7. If equals are subtracted from equals, the remainders are equal.
8. Things equal to the same thing are equal to each other.
9. Things that coincide are equal.
10. The whole is greater than any of its parts.

Now, we give informal statements of Hilbert's postulates for plane geometry.

Hilbert's Postulates (modified)

1. There are at least three noncollinear points in the plane.
2. Two points determine a unique line.
3. Every line contains at least two points.
4. If B is between A and C, then also B is between C and A, and the points are distinct and collinear.
5. If A, B, C are distinct points on a line, one and only one is between the other two.
6. If A and B are any two points, there exists a point C such that A is between B and C.
7. Each line separates the plane.
8. If $\overline{AB}$ is any segment and r is any ray from A', there exists a unique point B' on r such that $\overline{AB} = \overline{A'B'}$.

9. If $\overline{AB} \cong \overline{CD}$ and $\overline{EF} \cong \overline{CD}$, then $\overline{AB} \cong \overline{EF}$.

10. If B is between A and C, B' between A' and C' with $\overline{AB} \cong \overline{A'B'}$ and $\overline{BC} \cong \overline{B'C'}$, then $\overline{AC} \cong \overline{A'C'}$.

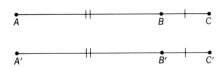

11. If $\angle A$ is any angle (not straight) and r any ray from A', then on a given side of r there is a unique ray s forming with r an angle, $\angle A'$, such that $\angle A \cong \angle A'$. Any two straight angles are congruent, and each angle is congruent to itself.

12. For triangles ABC and $A'B'C'$, if $\overline{AB} \cong \overline{A'B'}$, $\overline{BC} \cong \overline{B'C'}$, and $\angle B \cong \angle B'$, then $\angle C \cong \angle C'$.

13. If P is not on l, then there is not more than one line on P parallel to l.

We omit two postulates known as *Archimedes' postulate* and the *completeness postulate*. These two postulates enable one to establish a one-to-one correspondence between the set of real numbers and the points of any line.

Exercises 11–2

1. Relate Postulate 2 to one of Euclid's postulates.

2. Relate Postulate 3 to one of Euclid's.

3. Relate Postulate 10 to one of Euclid's.

4. Select two of Hilbert's postulates which have no counterpart in Euclid's postulates, and discuss how they correct omissions of Euclid.

5. Select two of Euclid's postulates which do not occur in Hilbert's set, and conjecture how these ideas will be treated in Hilbert's development.

6. Draw diagrams illustrating the meaning of each of Hilbert's Postulates 1 through 13, and show that these are all "intuitively obvious."

Perhaps you have noticed that Euclid's postulates do not mention numbers. At the time that the Greeks developed their geometry, algebra was almost nonexistent. In particular, number concepts were restricted to whole numbers and fractions. As you know, one must have irrational numbers in order to measure lengths of segments and areas of rectangles, and so Euclid developed his geometric theorems with no references to measurement. We mentioned that Hilbert's postulates enable one to establish a one-to-one correspondence between the set of real numbers and the points of any line. Several years ago mathematicians suggested that the postulates of geometry might be so phrased as to take advantage of real-number properties. In 1940, two Harvard mathematicians, G. D. Birkhoff and R. Beatley, wrote a high-school geometry text which utilizes properties of the real-number system. In this text (*Basic Geometry*, Scott, Foresman and Co.) only five postulates are listed.

Birkhoff–Beatley Postulates

1. The points on any line can be numbered so that number differences measure distances.
2. There is a unique line through two points.
3. All half-lines having the same endpoint can be numbered so that number distances measure angles.
4. All straight angles have the same measure.
5. Two triangles are similar if an angle of one equals an angle of the other and the sides including these angles are proportional.

Exercises 11–3

1. What connections can you find between the Birkhoff postulates and Euclid's? Hilbert's?
2. Draw diagrams illustrating these postulates.
3. Prove that each segment has a unique bisector.
4. Prove that each angle has a unique bisector.
5. Prove that two triangles are similar if two angles of one are congruent to two angles of the other.
6. Prove that base angles of an isosceles triangle are congruent and, conversely, that if two angles of a triangle are congruent, the sides opposite these angles are congruent.
7. Prove that two triangles are similar if their sides are proportional.
8. Prove that the angle sum for any triangle is 180°.

The Birkhoff–Beatley text was not well received. In 1960 a high-school text based on Hilbert's postulates was published (*Geometry*, Brumfiel, Eicholz, and Shanks, Addison-Wesley Publishing Company). In 1964 a geometry related to the Birkhoff-Beatley text was published (*Geometry*, Moise, Downs, Addison-Wesley Publishing Company). The postulates used in this latter text are much the same as those used in the School Mathematics Study Group experimental text.

We could list many other postulate sets for geometry. Currently there is considerable interest in transformational geometry. Texts are beginning to appear in which the axioms deal with properties of the basic rigid motions, reflections, rotations, and translations. It seems possible to conceive of an almost endless variety of postulational approaches to geometry.

12   Geometric Transformations

12–1   Functions

In the chapters on algebraic structures we considered functions (operations) upon sets of numbers, functions that added 4 to each real number, multiplied each real number by 3, etc. In this chapter we shall be concerned with functions which operate upon geometric figures, that is, upon sets of points rather than sets of numbers. Until now we have treated function ideas intuitively. It will be convenient to treat the concepts a bit more formally in this chapter.

The idea of a function is one of the central concepts of mathematics. Every function involves three factors.

a) A first set which we call the *domain* of the function.

b) A second set which includes a subset called the *range* of the function.

c) A *rule* that associates each element of the first set (domain) with an element of the range.

For example, let

$$X = \{1, 2, 3, 4\}, \qquad Y = \{1, 2, 3, 4, 5, 6, 7, 8, 9, 10\},$$

and let the rule be

Match each $x \in X$ *with* $x + 5 \in Y$.

If we denote this function by f, then

domain $f = X = \{1, 2, 3, 4\}$,

range $f = \{6, 7, 8, 9\} \subset Y$,

the rule $= \{(1, 6), (2, 7), (3, 8), (4, 9)\}$.

Sometimes the set of ordered pairs which we have designated as the rule is itself called the function. Note that this set of ordered pairs does give us all essential information about f. Explain how to extract domain f and range f from this set. Instead of listing this set of ordered pairs we can use the symbolism

$$\{(x, y) \mid x \in X \quad \text{and} \quad y = x + 5\}.$$

We read this as

The set of all ordered pairs x, y such that $x \in X$ and $y = x + 5$.

A technique for picturing functions is illustrated below.

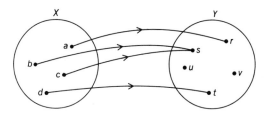

A function f from X into Y

Note that although the domain of the function (the set X) contains 4 elements the range has only 3 elements. We write $f(a) = r, f(b) = s, \ldots$, and read this as "$f$ of a equals r," "f of b equals s," ....

The sets X and Y may be any sets imaginable. In everyday life we deal with many functions. Language which usually signals concern with a function is "the ... of ...." Examples are

a) *The* mother *of* ...

b) *The* height *of* ...

c) *The* age *of* ...

d) *The* responsibilities *of* ...

In (a) above, the domain and range are both sets of people. In (b)and(c) the range is a set of numbers. In (d), each element of the range is itself a *set*, a set of responsibilities.

The notation

$$f: X \to Y; \quad X \xrightarrow{f} Y$$

indicates that we are considering a function named f which matches each element of a set X with an element of a set Y. We say that

f is a function *from X into Y.*

A function from X into Y is *one-to-one* (1–1) if it matches different elements of X with different elements of Y. The functions *the mother of* and *the age of* are not 1–1. Two people can have the same mother or can be the same age. But our numerical function above which matches each $x \in \{1, 2, 3, 4\}$ with $x + 5$ is 1–1. (Show that this is so.)

A function from X *into* Y is said to be an *onto* function if each element of Y is paired with some element of X. In our numerical example, f is not an onto function. The number 3 in Y has no mate in X. On the other hand, the function g that matches each whole number x with the even number $2x$ is an onto function from W to E, where E is the set of all even numbers. Note that g is also a 1–1 function,

$$g: W \xrightarrow[\text{onto}]{\text{1-1}} E; \qquad \forall x \in W, \ g(x) = 2x.$$

Functions which are both 1–1 and onto are of particular importance. These functions are ordinarily called *one-to-one correspondences*. The ideas of *congruence*, *similarity*, and *symmetry* can be interpreted as 1–1 correspondences which match a first set of points 1–1 with a second set.

The diagram below shows that when we reverse the arrows which picture a 1–1 correspondence from X to Y, we picture a new 1–1 correspondence from Y to X. We call each function the *inverse* of the other. In our diagram we let f be the function that adds 3 to each element in its domain. Then the inverse function, denoted by f^{-1}, subtracts 3. (Read f^{-1} as "f inverse".)

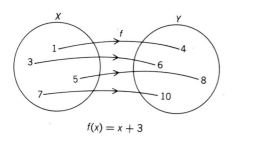

$f(x) = x + 3$

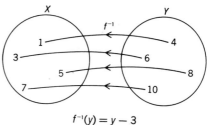

$f^{-1}(y) = y - 3$

Exercises 12–1

1. When a child counts the elements of a set, we expect him to construct a function *from* a set of number sounds {one, two, three, ...} *to* the set of objects. We expect this function to be 1–1 and onto. We also expect the child to select his number sounds in a certain approved order. The following diagrams indicate how a small child might touch objects and "count." Explain the criticisms of his efforts, and comment upon the assertion that a child must understand a great deal about functions before he can count properly.

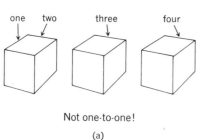

Not one-to-one!

(a)

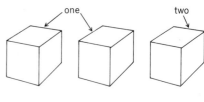

Not a function!

(b)

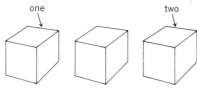

Not onto!

(c)

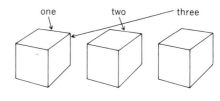

Neither one-to-one nor onto!

(d)

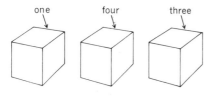

The domain is the wrong set!

(e)

2. Select a book, newspaper, or magazine article and read part of it, identifying references to functions associated with the terminology, "the ... of ...". Describe domains and ranges for several of these functions.

3. When negative numbers are introduced, it is customary to invent for each $a \in W$ the number ^-a, called *minus a*. The symbol "$-$" denotes a function. If we denote the set $\{^-0 = 0,\ ^-1,\ ^-2, \ldots\}$ by W^-, then

$$-: W \xrightarrow[\text{onto}]{\text{1-1}} W^-.$$

Describe how this one-to-one correspondence between W and W^- is used to construct a function which maps $W \cup W^-$ one-to-one onto itself. What is this function called?

4. Let $f: W \xrightarrow{\text{into}} O$, where W is the set of whole numbers, $\{0, 1, 2, \ldots\}$ and O is the set of odd whole numbers.

  a) Given that $f(x) = 2x + 1$; show that f is 1–1 and onto.

  b) Given that $f(x) = 2x + 7$; show that f is 1–1 but not onto.

  c) Given that $f(x) = x + 1$ if x is even and $f(x) = x + 4$ if x is odd; show that f is onto but is not 1–1.

5. Let $f: I \to W$, where $f(x) = x^2$, I is the set of integers, and W the set of whole numbers. Show that f is neither 1–1 nor onto.

6. The geometric diagrams below suggest 1–1 onto functions from a first set of points to a second set. Describe these functions. Note the inverse of each function.

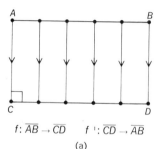

$f: \overline{AB} \to \overline{CD}$ $f^{-1}: \overline{CD} \to \overline{AB}$

(a)

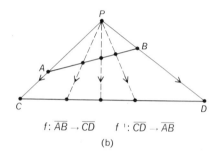

$f: \overline{AB} \to \overline{CD}$ $f^{-1}: \overline{CD} \to \overline{AB}$

(b)

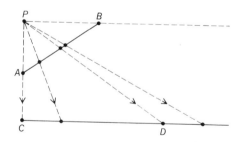

Let X be the set consisting of all points of $\overline{AB}$ except point B. Line $\overleftrightarrow{PB}$ is parallel to ray $\overrightarrow{CD}$.

$f: X \xrightarrow[\text{onto}]{1\text{-}1} \overrightarrow{CD}$ $f^{-1}: \overrightarrow{CD} \xrightarrow[\text{onto}]{1\text{-}1} X$

(c)

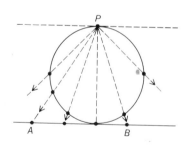

Let Y be the set consisting of all points of the circle except P.

$f: Y \xrightarrow[\text{onto}]{1\text{-}1} \overleftrightarrow{AB}$ $f^{-1}: \overleftrightarrow{AB} \xrightarrow[\text{onto}]{1\text{-}1} Y$

(d)

7. Describe how one might construct the following:

  a) A 1–1 correspondence between two congruent segments.

  b) A 1–1 correspondence between any two segments, congruent or not.

c) A 1–1 correspondence between any two triangles.

d) A 1–1 correspondence between any two circles; between any circle and any convex polygon; between any circle and any simple polygon, convex or not; between any circle and any simple closed curve.

Diagrams of the type we have used in Exercises 6 and 7 can be used to portray the important concept, *composition* of functions.

The figure below pictures a function f from $\overline{AB}$ to $\overline{CD}$, a function g from $\overline{CD}$ to $\overline{EF}$, and the composite of these two functions, $g \circ f$ [also written $g(f)$, gf, and g of f] from AB to EF.

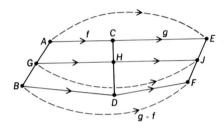

$$f(A) = C; \quad g(C) = E; \quad g(f(A)) = g \circ f(A) = E$$

We comment briefly on functional notation. The usual $f(x)$ notation has the following interpretation:

a) The symbol f names the function.

b) The symbol x names an element upon which the function "acts."

c) The complete symbol $f(x)$ names the element that the function f matches with x.

For example, if f is the function that adds 4 to each whole number, then the symbol $f(5)$ is just a name for 9.

We often use function notation in other styles. For example, the notation that we use to name segments, rays, lines, etc., is actually function notation.

$\overset{\bullet}{B}$

$\overset{\bullet}{A}$ $\overset{\bullet}{C}$

$\overset{\bullet}{D}$

For example, let $A, B, C, D, \ldots$ denote points. In the symbol $\overset{\leftrightarrow}{AB}$, the symbol AB denotes the ordered pair of points (A, B), and the symbol $\leftrightarrow$ denotes the "line function" which matches this ordered pair of points with a line. The full symbol $\overset{\leftrightarrow}{AB}$ denotes the line that is matched with this pair of points. Point out, similarly, the use of function notation in the symbolism

$$\triangle ABC, \ \angle RST, \ \overline{AB}, \ \sqrt{15}, \ x^2.$$

12–2 Motions

Geometric transformations are often called *motions*, because it is natural to think of the transformations as moving points from one position to another. There are many physical motivations for this point of view. For example, we can imagine a collection of floating particles on a river. At one instant they occupy particular positions. A moment later they have been moved by the current to new positions. In a physical example like this we may be interested in the *path* that each particle traverses in moving from its old position to its new position. However, when we consider geometric transformations as motions we usually pay little attention to paths and consider only the initial and final positions of points.

Naturally, we are primarily interested in those geometric transformations which have rather simple descriptions. Some of the terms used to describe basic transformations are *translations, rotations, reflections, enlargements,* and *projections.*

In an earlier chapter we saw that we could think of linear functions as translations, rotations, or stretchings of a line. For example, if $L(x) = x - 3$, L is a *translation*; if $L(x) = -x$, L is a 180° *rotation* of the line about the origin. If $L(x) = 3x - 12$, then L leaves the point 6 fixed, but it *stretches* the line so that all distances are multiplied by the factor 3. In general, if $L(x) = ax + b$ and a is neither 1 nor -1 we shall call L an *enlargement.*

As you have seen, concepts of algebra and geometry are brought together through the consideration of transformations. Of course, in geometry we are interested principally in transformations of a plane or of space, rather than of a line. It is easy to extend the notions of transformations that we have mentioned here to planes and space. The diagram below pictures a translation which maps a *plane* upon itself. The vector **v** indicates how every point in the plane is moved by the translation T. Each point of α is moved in the direction of **v** a distance equal to the length of **v**.

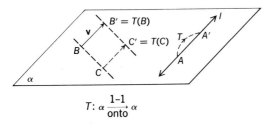

Exercises 12–2

1. For the translation T pictured above, note that

$$T\colon l \xrightarrow[\text{onto}]{1\text{-}1} l.$$

That is, line l is translated onto itself.

a) What other lines of the plane have this property?

b) Is any point of the plane mapped upon itself?

2. Use your knowledge of the geometry of parallelograms to show that the line $\overleftrightarrow{B'C'} = T(\overleftrightarrow{BC})$ is parallel to $\overleftrightarrow{BC}$. Hence if m is any line in the plane α, what can you say about $T(m)$?

3. Use your knowledge of parallelograms to prove that the segment $\overline{B'C'}$ has the same length as has $\overline{BC}$. If $\overline{XY}$ is any segment in α, what can you say about the segment $T(\overline{XY})$?

4. The transformation T maps each triangle, $\triangle ABC$, into a triangle, $\triangle A'B'C'$. How are these triangles related?

5. We can think of the translation T as moving every point in three-dimensional space.

a) Describe the lines in space that are mapped upon themselves.

b) Describe the planes in space that are mapped upon themselves.

6. What vector determines the inverse of the translation T?

When we visualized M_{-1} as a rotation of a line, we saw it as a 180° rotation about the origin. In the plane we can consider rotations of any angle about any point. We choose a *center* C, an *angle* of rotation, and a *sense* of rotation (clockwise or counterclockwise).

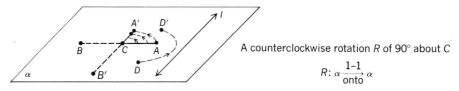

A counterclockwise rotation R of 90° about C

$$R\colon \alpha \xrightarrow[\text{onto}]{1\text{-}1} \alpha$$

Exercises 12–3

1. Consider the rotation R pictured above. Is any point of α other than C mapped upon itself? Are any line segments mapped upon themselves? Any triangles? Any circles? Any squares?

2. Let l be any line of the plane. How are the lines l and $l' = R(l)$ related?

3. Describe the inverse of the transformation R.

4. Describe the composite functions for the following:

   a) $RR = R^2$,   b) $RRR = R^3$,   c) R^4.

   [We shall write RR instead of $R \circ R$ or $R(R)$.]

5. Is the 180° counterclockwise rotation about C exactly the same function as the 180° clockwise rotation about C?

6. The rotations of the plane α about point C form a group of transformations. Verify the calculations below. The symbol R_{30} denotes the 30° counterclockwise rotation about C.

   a) $R_{200}R_{180} = R_{20}$.

   b) $R_{90}R_{270} = R_0$.

   c) $R_{120}^4 = R_{120}$.

7. If P is any point of the plane other than C, then by repeatedly applying the 90° rotation R pictured above, P can be carried to any one of 4 points:

   $R(P),$ $R^2(P),$ $R^3(P),$ $R^4(P) = P.$

   a) Describe the quadrilateral which has these points for vertices.

   b) If l is any line not on C, describe the quadrilateral formed by the four lines l, $R(l)$, $R^2(l)$, and $R^3(l)$.

8. Answer questions analogous to those of Exercise 7 for a 60° rotation R_{60}, a 10° rotation R_{10}, and a 17° rotation R_{17}.

9. How would you generalize the concept of a rotation in a plane to a rotation in three-dimensional space?

In the plane we can consider *reflections* in both points and lines. For a reflection in a point C, we match each point P with the point P' so that C is the midpoint of $\overline{PP'}$. But this is simply the 180° rotation about C.

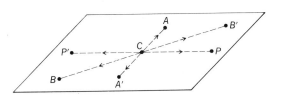

The *reflection* in C is the 180° rotation about C

The reflection in a line l matches every point P on one side of l with the point P' on the other side, such that l is the perpendicular bisector of $\overline{PP'}$. Each point of l remains fixed. You can think of l as a mirror. We denote reflections by M (for mirror). When we speak of a reflection of a plane, we shall always mean a reflection in a line rather than a point.

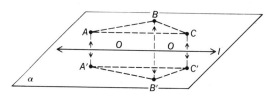

The *reflection* M_l in line l

$$M_l\colon \alpha \xrightarrow[\text{onto}]{\text{1-1}} \alpha$$

It is clear that every reflection is its own inverse. That is, if $M_l(A) = A'$, then $M_l(A') = A$. If we denote the identity transformation that leaves every point of the plane fixed by I, then

$$M_l^2 = M_l M_l = I.$$

One special property of reflections should be noted. We say that each reflection "changes the orientation" of the plane. The diagram above illustrates this. Note that the vertices A, B, C of $\triangle ABC$ are arranged in *clockwise* order. But the corresponding vertices A', B', C' of the image triangle, $\triangle A'B'C'$, are in *counterclockwise* order. You are familiar with the way a mirror interchanges "right" and "left."

Exercises 12–4

1. Verify for each of the following triangles that the reflection M_l pictured above changes the orientation of the vertices.

    a) $\triangle ABC'$,   b) $\triangle AB'C$,   c) $\triangle A'CC'$.

2. It is "obvious" that if $\overline{AB}$ is any segment and $\overline{A'B'}$ is the mirror image of $\overline{AB}$ in a line l, then the lengths of the segments are the same. Supply the reasons for the formal proof of this for the case illustrated below, where A and B are on the same side of l. Point P is chosen on l so A, B, P are not collinear.

    a) $\triangle AMP \cong \triangle A'MP$. (Why?)

    b) $\triangle BNP \cong \triangle B'NP$. (Why?)

    c) $\angle APB \cong \angle A'PB'$. (Why?)

d) $\triangle APB \cong \triangle A'PB'$. (Why?)

e) $\overline{AB} \cong \overline{A'B'}$. (Why?)

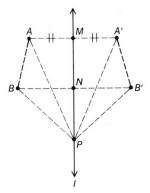

3. Explain how the reflection of a line in a point changes the orientation of the line.

4. Sketch a diagram picturing a reflection of three-dimensional space in a line. Relate this transformation to rotations in space about an axis.

5. Sketch a diagram picturing a reflection of three-dimensional space in a plane.

6. Consider the reflection M_l of a plane α.

a) Given that $m \parallel l$; describe the line $M_l(m)$.

b) Given that $n \perp l$; describe the line $M_l(n)$.

c) Given that p intersects l at A; describe the line $M_l(p)$.

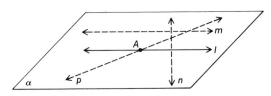

7. If A and B are two points in a plane α, then there is exactly one line l in α such that $M_l(A) = B$, and $M_l(B) = A$. Describe this line l.

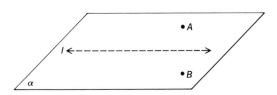

8. If $\angle ABC$ is any angle in a plane α, then there is exactly one line l in α such that

$$M_l(\vec{BA}) = \vec{BC} \quad \text{and} \quad M_l(\vec{BC}) = \vec{BA}.$$

Describe this line l.

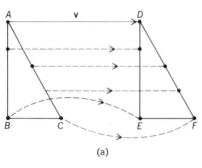

The three types of motions, translations, rotations, and reflections, are the basic *rigid motions* of Euclidean geometry. These generate the *congruence* transformations that do not affect size or shape of geometric figures. We say that two figures are *congruent* if one can be mapped upon the other by a sequence of these basic transformations. The diagrams below indicate how one triangle may be mapped upon another in various ways by means of congruence functions.

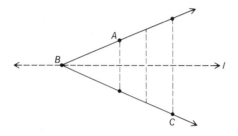

$\triangle ABC \cong \triangle DEF$ because the translation T determined by the vector $\mathbf{v}$ maps $\triangle ABC$ upon $\triangle DEF$.

(a)

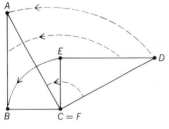

$\triangle DEF \cong \triangle ABC$ because the 90° counterclockwise rotation R with center C maps $\triangle DEF$ upon $\triangle ABC$.

(b)

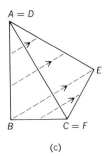

A = D

B   C = F

(c)

$\triangle ABC \cong \triangle DEF$ because the reflection M
in $\overleftrightarrow{AC}$ maps $\triangle ABC$ upon $\triangle DEF$.

It may be necessary to use a combination of these basic transformations to construct a congruence from one figure to a second. For each figure below, a sequence of transformations is suggested. Describe each sequence.

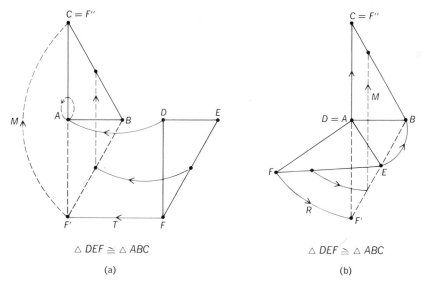

$\triangle DEF \cong \triangle ABC$

(a)

$\triangle DEF \cong \triangle ABC$

(b)

Often in school geometry a "proof" of the side-angle-side congruence theorem for triangles is presented by arguing that one triangle can be "moved" so that it "coincides" with the other. Recall this proof, and relate it to the transformations we have discussed. Why is a proof of the side-side-side congruence theorem for triangles not so easily effected by these methods?

Enlargements in the plane (and in space) change the sizes of figures but not their shapes. Each enlargement E is determined by a point (center) and nonzero real number (stretching factor). If we denote the center by C and the real number by r, then if $r > 0$, each point X is carried to a point $E(X) = X'$ *on the ray* $\overrightarrow{CX}$ such that

$$m(\overline{CX'}) = r \cdot m(\overline{CX}).$$

If $r < 0$, the image point X' is on the ray opposite $\overrightarrow{CX}$. We picture below three enlargements about points A, B, C with stretching factors $\frac{1}{2}, 2$, and $-\frac{1}{2}$, respectively. Each may be considered to be a transformation of the plane or of space. Note that enlargements determined by negative real numbers change the orientation of the plane (or space). It is intuitively obvious that an enlargement maps each geometric figure upon a figure similar to it. (Can you use your knowledge of similar triangles to prove that an enlargement maps each triangle upon one similar to it?)

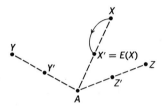

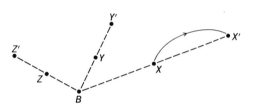

E maps each point X onto the
point X' *halfway* from A to X.

$E(X)$ is on $\overline{BX}$ and *twice*
as far from B as is X.

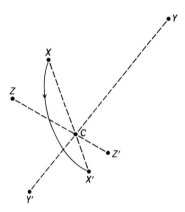

$E(X)$ is on the ray opposite $\overrightarrow{CX}$ and
half as far from C as is X.

The transformations known as *projections* belong to the subject matter of affine and projective geometry. These transformations distort both size and shape. We shall do little more than define *parallel* and *central* projections.

If we restrict our attention to one plane, then these projective transformations, the parallel and central projections, map points of one line upon points of a second. We picture next a parallel projection P that maps the points of a line l upon the points of a line m. The diagram is self-explanatory. We pick the family of all lines in a plane parallel to a fixed line, and then the lines of this family map the points of any transversal l upon the points of any other transversal m, as shown above. Note that if l and m intersect, the intersection point is mapped upon itself. Note also that if $\overline{AB}$ is mapped upon a segment $\overline{A'B'}$ twice as long as $\overline{AB}$, then every segment on l is mapped upon a segment twice as long. (Why?) Describe the inverse of this parallel projection.

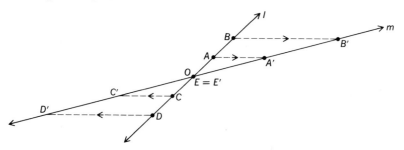

A parallel projection $P: l \xrightarrow[\text{onto}]{1\text{-}1} m$

By forming the composite of two parallel projections we can map a line upon itself. We illustrate this below. If we denote by C the point of intersection of l and m, then the composite of these two parallel projections, P_2P_1, is an *enlargement* of l with center C. Show that this is the case.

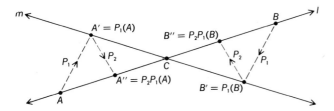

Central projections are the basic transformations of projective geometry. These transformations "mix up" points of a line much more than the other transformations we have considered. Again we restrict ourselves to a single plane and will think of each central projection as a function mapping the points of one line l upon a second line m. The diagram at the top of page 266 makes the ideas clear. Describe the inverse of the projection pictured.

Central projections distort distances. (Compare $\overline{AB}$ with $\overline{A'B'}$.) In particular, note that betweenness relations may be destroyed by a central projection. In the figure above, B is between points A and D on line l. When you examine the image points A', B', and D' on line m, what do you observe? Note also that there is one point E on l unmatched with any point of m. Also, there is a point F' of m which is not the image of any point of l. In the development of projective geometry a new

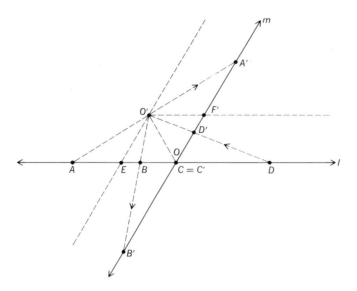

point is invented "at infinity" on each line. *E* is matched with the point at "infinity" on *m*, and *F'* is matched with the point at "infinity" on *l*. These matters need not concern us here. By composing two central projections, we can map a line upon itself. Sketch a diagram illustrating this.

When we look at parallel and central projections in space we see them as transformations which map all the points of one plane upon the points of a second. The following two diagrams illustrate these transformations. Explain them.

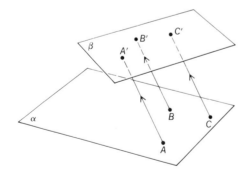

A parallel projection $P: \alpha \to \beta$

A central projection that maps points of a plane upon a parallel plane does not distort shape, but it does distort size. (Justify this claim.) What have central projections to do with photography?

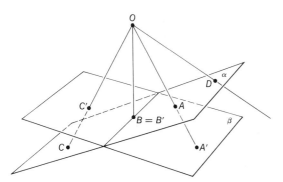

A central projection $P: \alpha \to \beta$

Exercises 12–5

1. Describe each transformation pictured below as a single translation, rotation, or reflection. For translations, describe the direction and distance of translation; for rotations, describe center and angle of rotation; for reflections, indicate the line that determines the reflection.

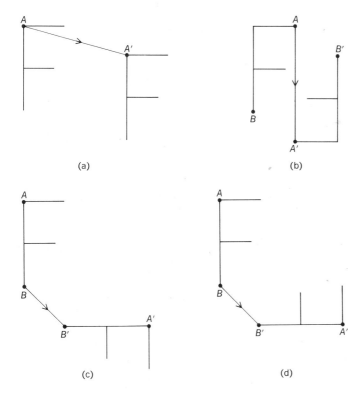

(a)   (b)

(c)   (d)

2. In this exercise assume that all five segments which form the letter *H* are of equal length. Describe the transformation suggested by each diagram.

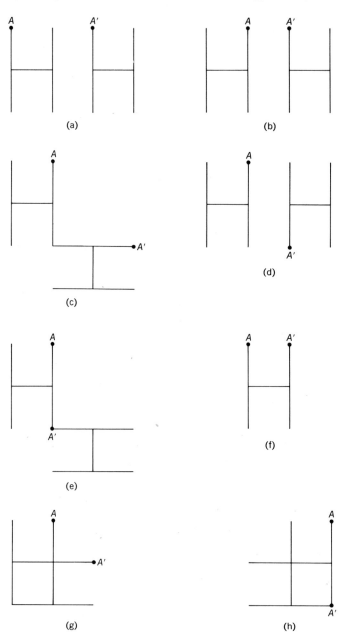

3. Some letters can be mapped upon themselves by rotations or reflections. Consider each capital letter and describe all transformations that map that letter upon itself. The set of all transformations that map a particular letter upon itself (including the identity) constitute a *group* of transformations. For each letter specify the number of elements in its group. Answers are given for a few letters. Note that F has only the identity transformation.

A   B   C   D   E   F   G   H   I   J   K   L   M
↓   ↓
(2)   (1)

N   O   P   Q   R   S   T   U   V   W   X   Y   Z
    ↓   ↓
(infinitely   (4)
many)

4. Each of the small letters p, b, d can be transformed into either of the other two by transformations of the kind we have considered. Describe these transformations.

5. Consider two points A and B on a line l.

   a) Describe a translation T, a reflection M, and an enlargement E, each of which maps A upon B.

   b) Describe an enlargement that maps A upon B and B upon the midpoint of $\overline{AB}$.

6. Copy the figure below and picture two parallel projections,

   $P_1 : l \to m$ and $P_2 : m \to l$,

   such that

   $P_2 P_1 : l \to l$ and $P_2 P_1 (A) = B$.

   What is $P_2 P_1 (B)$?

   What is $P_2 P_1 (C)$?

   Describe $P_2 P_1$ as a translation.

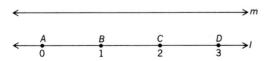

7. Copy the figure below and picture two parallel projections,

$$P_1: l \to m \qquad \text{and} \qquad P_2: m \to l,$$

such that

$$P_2P_1: l \to l \qquad \text{and} \qquad P_2P_1(A) = B.$$

What is $P_2P_1(B)$?

What is $P_2P_1(C)$?

Describe P_2P_1 as an enlargement.

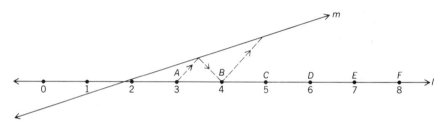

8. Copy the figure below and picture three parallel projections,

$$P_1: l \to m, \qquad P_2: m \to n, \qquad P_3: n \to l,$$

such that

$$P_3P_2P_1: l \to l \qquad \text{and} \qquad P_3P_2P_1(A) = A, \ P_3P_2P_1(B) = C.$$

Show that $P_3P_2P_1(C) = E$ and $P_3P_2P_1(D) = G$. Show that $P_3P_2P_1$ is described by the equation $P_3P_2P_1(x) = 2x + 1$.

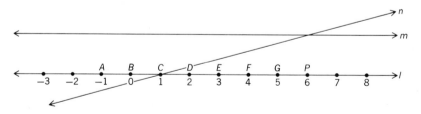

9. The following figures picture two transformations, a parallel projection P and a central projection C. In each case the composite transformations CP and PC can be viewed as mapping line l upon itself. Equations are given for these composite transformations for each of (a) and (b). Verify that the equations seem to be correct and attempt to compute similar equations for parts (c) and (d).

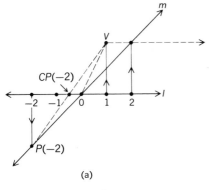

(a)

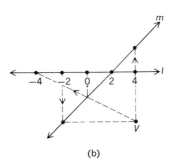

(b)

$$CP(-2) = -\frac{1}{2} \quad CP(x) = \frac{x}{2-x}$$

$$PC(2) = \frac{4}{3} \quad PC(x) = \frac{2x}{x+1}$$

$$CP(x) = \frac{8x-8}{x+2}$$

$$PC(x) = \frac{-2x-8}{x-8} = \frac{2x+8}{8-x}$$

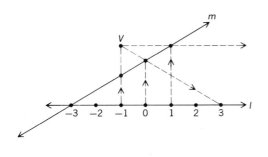

(c)

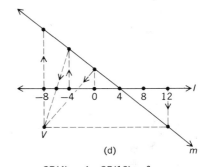

(d)

$$CP(0) = 3; \quad CP(-1) = -1$$
$$CP(1) = \infty; \quad CP(-3) = ?$$
$$CP(2) = ?; \quad CP(-2) = ?$$

$$CP(4) = 4; \quad CP(12) = ?$$
$$CP(-8) = ?; \quad CP(-4) = ?$$

(i) $CP(x) = ?$ (ii) $PC(x) = ?$ (i) $CP(x) = ?$ (ii) $PC(x) = ?$

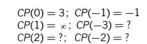

$$\frac{5x+3}{1-x} \qquad \frac{x-3}{x+5} \qquad \frac{-16x+32}{x-12} \qquad \frac{32x+96}{3x+44}$$

12–3   The Group of Rigid Motions

The set of rigid motions of the plane consists of translations, rotations about a point, reflections in a line, and all possible combinations of motions of this type. These are the transformations of the plane that preserve congruence. They are called *congruences* or *isometries*. The set is clearly a group relative to the operation of composition. The operation is associative. (Why?) There is an identity transformation. (What is it?) And every transformation has an inverse. (Explain.)

In this section we make a brief analysis of this group. It turns out that the structure of the group of rigid motions is best explained in terms of reflections. It is a surprising fact that every translation and every rotation is the product of two reflections. The diagrams below indicate how this can be proved.

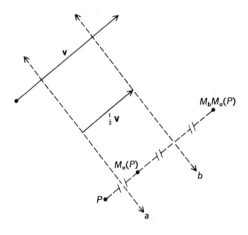

Let the vector $\mathbf{v}$ determine a translation T. Let a, b be any two lines perpendicular to $\mathbf{v}$ and such that a vector from a to b and perpendicular to these lines is half as long as $\mathbf{v}$. Then $M_b M_a = T$. (What is $M_a M_b$?)

Let C be the center and θ the counterclockwise angle of a rotation R. Let a, b be any two lines on C such that the counterclockwise angle from a to b is $\frac{1}{2}\theta$. Then $M_b M_a = R$. (What is $M_a M_b$?)

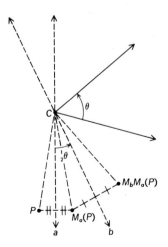

The reader should draw several figures, try several different positions for points, and satisfy himself that the above assertions are correct.

We see, then, that the group of rigid motions in the plane can be viewed as just the group formed by reflections in lines. In the next list of exercises we will see how this group can be completely analyzed and described. The major result is that the product of any number of reflections can be expressed either as a single reflection, or as the product of two reflections (and so is a translation or rotation), or as the product of three reflections with two of the three lines parallel and the third perpendicular to them.

Exercises 12–6

1. Given that $a \perp b$. Show that $M_b M_a = M_a M_b$.

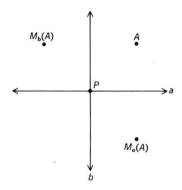

2. Lines a, b, c, d are concurrent, and the angle from a to b has the same measure and sense as has the angle from c to d. Show that

$$M_b M_a = M_d M_c.$$

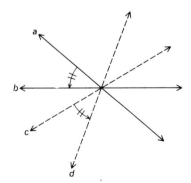

3. Lines a, b, c are parallel. Show that $M_c M_b M_a = M_d$, where $d \parallel c$ and lines d, c are in the same relative position as·are lines a, b.

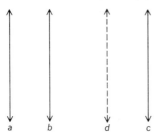

4. Given that $a \parallel b$ and $c \perp a$. Show that

$$M_a M_b M_c = M_a M_c M_b = M_c M_a M_b,$$

and that

$$M_b M_a M_c = M_b M_c M_a = M_c M_b M_a.$$

Show that the result of each of these reflections is simply a reflection in c followed by a translation parallel to c. This transformation is called a *glide reflection*.

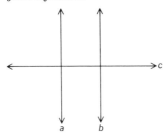

5. Use the result of Exercises 1, 2, and 4 to show that the composite of three reflections, M_c, M_b, M_a, in lines a, b, c which are not all parallel to one another, is a glide reflection. The diagram below suggests the argument.

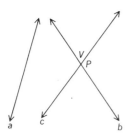
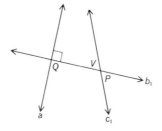
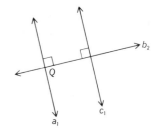

$$M_c M_b M_a = M_{c_1} M_{b_1} M_a = M_{c_1} M_{b_2} M_{a_1}$$

$$\angle c_1, b_1 \cong \angle c_1, b_2$$

6. Given that $a \parallel b$, $c \parallel d$, and $a \not\parallel c$. Show how to construct lines e and f such that $e \parallel f$ and $M_a M_c M_b M_a = M_f M_e$.

7. Use the results of the preceding exercises to show that no matter how many reflections may be performed successively in lines $a_1, a_2, \ldots, a_n$, the composite of all these reflections is either

   a) A reflection in one line,

   b) The composite of two reflections and hence a translation or a rotation, or

   c) The composite of three reflections with two lines parallel and the third perpendicular to these two, hence a glide reflection.

8. Images $F_{(1)}, F_{(2)}, \ldots, F_{(6)}$ of the centered letter F are pictured below. For each image F_i, describe the rigid motion that maps F upon F_i, and place it in the correct category of Exercise 7 above.

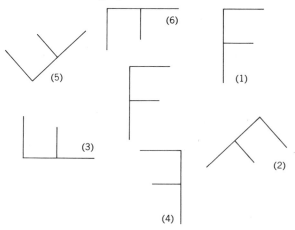

9. Suppose that $M_a M_b \neq I$, but $(M_a M_b)^2 = I$. What can you say about the lines a and b? [I is the identity transformation.]

10. Suppose that $M_a M_b \neq I$, but $(M_a M_b)^5 = I$.

    a) Prove that $M_a M_b$ is a rotation.

    b) Show that there are two possibilities for the smallest angle from b to a.

11. In the figure on the next page the squares are numbered in spiral fashion. The lines containing the sides of the square labeled 1 are named a, b, c, and d. Note that M_a: $1 \to 4$; M_b: $1 \to 6$; $M_b M_a$: $1 \to 5$. That is, the reflection M_a maps the points of square 1 and its interior onto corresponding points of square 4, the rotation $M_b M_a$ maps square 1 upon square 5, etc. Imagine this numbering continued over the entire plane and compute:

    a) $M_c M_a$ (1)

    b) $M_a M_c$ (1)

c) $M_a M_b M_c$ (1)

d) $M_c M_b M_a$ (1)

e) $(M_a M_c)^2$ (1)

f) $(M_a M_c)^3$ (1)

g) $(M_a M_b M_c M_d)^2$ (1)

h) $(M_a M_b M_c)^2$ (1)

	b		d					
65	64	63	62	61	60	59	58	57
66	37	36	35	34	33	32	31	56
67	38	17	16	15	14	13	30	55
68	39	18	5	4	3	12	29	54
69	40	19	6	1	2	11	28	53
70	41	20	7	8	9	10	27	52
71	42	21	22	23	24	25	26	51
72	43	44	45	46	47	48	49	50
73	74	75	76	77	78	79	80	81

(lines a and c point to the left at the left edge of the grid)

12. The questions below refer to the figure of Exercise 11.

a) If only reflections in lines a, b, c, and d are employed, do you think it possible to map square 1 upon every square of the plane? Justify your answer.

b) Find the smallest number of reflections, using only M_a, M_b, M_c, and M_d, needed to map square 1 upon square 73.

c) Suppose that you have found one product of reflections (in lines a, b, c, and d) which maps square 1 upon square 213 and a second product mapping square 1 upon square 475. Explain how to construct a rigid motion that will map square 213 on square 475.

d) Verify for several cases that square 1 can be mapped upon square n^2 by $n - 1$ reflections selected from M_a, M_b, M_c, and M_d.

13. Consider the network of lines partitioning the plane into congruent equilateral triangles with the triangles systematically numbered as shown in the accom-

panying figure. Lines a, b, c form the sides of the triangle labeled 1. We consider only reflections in these lines. Imagine the numbering continued over the plane.

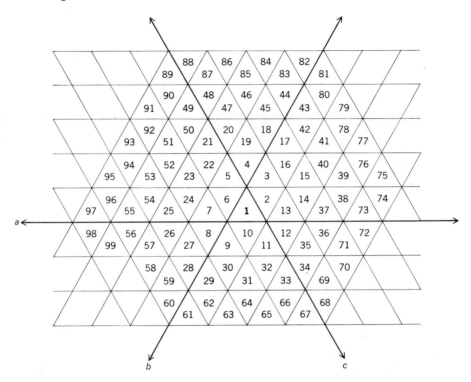

a) Compute: $M_a(1)$; $M_b(1)$; $M_c(1)$; $M_a M_b(1)$; $M_b M_a(1)$; $(M_a M_b)^3(1)$.

b) Show that $M_b M_c M_b = M_c M_b M_c$, and that this transformation is the reflection in the line parallel to a and concurrent with b and c.

c) Show how to obtain the reflection in the line parallel to b, and concurrent with a and c; parallel to c, and concurrent with a and b.

d) Note that the lines a, b, c together with the three lines mentioned in parts (b) and (c) above contain all the sides of triangles 1, 2, 6, and 10. Show how to obtain as products of M_a, M_b, and M_c the reflections in all lines which contain a side of triangle 5, 7, 8, 11, 13, or 3.

e) Argue that the reflection in every line in the infinite network above can be expressed as a product of the reflections M_a, M_b, M_c. For example, the reflection in the line separating triangles 7 and 24 is

$$(M_b M_c M_b)(M_a M_b M_a)(M_b M_c M_b).$$

f) How does each reflection M_a, M_b, M_c affect even-numbered triangles? Odd-numbered triangles?

g) Note that triangles 1 through 6 form a hexagon as do triangles 1 through 24 and also triangles 1 through 54. Can you describe a pattern?

12–4 Groups of Symmetries

In an earlier chapter we treated symmetries of several geometric figures algebraically as permutations of their vertices. In this section we shall look at some of the same symmetric figures, but our viewpoint will be more geometric. We shall use what we know about reflections, relying mainly upon the fact that the product of two reflections in intersecting lines is a rotation about the point of intersection through an angle which is double the angle from the first line to the second. From this point of view, look again at the symmetries of a rectangle.

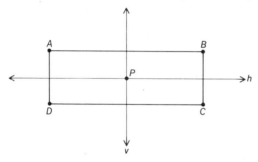

Since lines h and v are perpendicular, the product $M_h M_v$ is a 180° counterclockwise rotation about P. Interpret $M_v M_h$ similarly, and explain why $M_v M_h = M_h M_v$.

For the equilateral triangle pictured below, all symmetries are generated by the reflections M_a, M_b, M_c in the altitudes a, b, c which intersect at the center of gravity G. As the diagram indicates, we can think of the angle from a to b as a

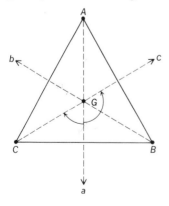

60° clockwise angle or as a 120° counterclockwise angle. Hence, we can interpret $M_a M_b$ either as a 120° clockwise rotation or a 240° counterclockwise rotation. Give similar interpretations for $M_a M_c$, $M_c M_a$, $M_c M_b$, $M_b M_c$.

The exercises below present other results that we obtain by viewing symmetries as composed from reflections.

Exercises 12–7

1. For the symmetries of the equilateral triangle pictured on page 278, we have noted that

    i) $M_a M_b = M_b M_c = M_c M_a$;   each is a 120° clockwise rotation.

    ii) $M_b M_a = M_c M_b = M_a M_c$;   each is a 120° counterclockwise rotation.

    Of course $M_a^2 = M_b^2 = M_c^2 = I$. Use these results to show that

    a) $M_a M_b M_c = M_b$,   b) $M_b M_c M_a = M_c$,

    c) $M_c M_a M_b = M_a$,   d) $M_c M_b M_a = M_b$.

2. For the square pictured below, all the symmetries can be generated by reflections in one of the diagonals and either h or v.

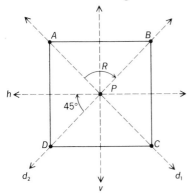

    a) Given that R is the 90° clockwise rotation about P. Explain why

    $R = M_h M_{d_2}$ and $R^3 = M_{d_2} M_h$.

    b) Show that $M_h M_{d_2} = M_{d_2} M_v = M_v M_{d_1} = M_{d_1} M_h$.

    c) Use the result of (b) to show that

    $M_{d_2} M_h M_{d_2} = M_v$ and $M_{d_2} M_h M_{d_2} M_h M_{d_2} = M_{d_1}$.

    d) Now justify the assertion that the group of all symmetries of the square is generated by M_{d_2} and M_h.

3. The square and equilateral triangle are special regular polygons. The group of symmetries of any regular polygon can be analyzed in a similar fashion. For

each of the regular polygons below explain how the group of all symmetries can be generated by the two indicated reflections.

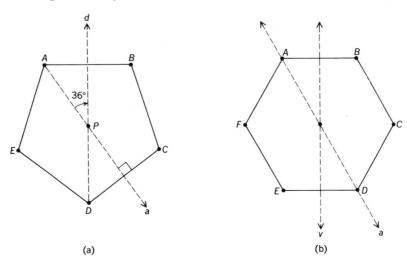

(a)   (b)

a) Show that $M_d M_a$ is a 72° rotation about P, and that these two reflections generate all ten symmetries of the regular pentagon. Describe these symmetries geometrically.

b) Show that M_v and M_a generate all twelve symmetries of a regular hexagon. Describe these symmetries geometrically.

4. If we return to the algebraic notation we used when we presented symmetries as permutations of the vertices of a figure, then for the square pictured in Exercise 2 we have

$$M_h = \begin{pmatrix} A & B & C & D \\ D & C & B & A \end{pmatrix} \quad \text{and} \quad M_{d_2} = \begin{pmatrix} A & B & C & D \\ C & B & A & D \end{pmatrix}.$$

Use the technique for multiplying permutations, and verify that M_v and M_{d_1} can be expressed as products of these symmetries. Treat part (a) or (b) of Exercise 3 similarly.

12–5 Transformations and Proofs

In this section we shall use transformations in the proofs of several geometric theorems. It should be emphasized that our approach to these proofs is intuitive. We take many things for granted that would require proof in a careful axiomatic development. For example, we assume as obvious that the rigid motions (reflections, rotations, and translations) have the following properties.

1. Every rigid motion maps every geometric figure S onto a figure S' such that $S \cong S'$.

2. Every segment is mapped upon itself by the reflection in its perpendicular bisector, and every angle is mapped upon itself by the reflection in its bisector.

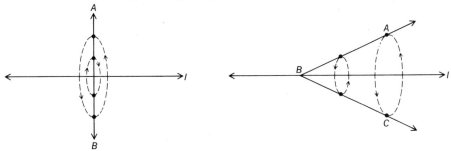

3. Each translation carries each line l into a line l' such that either $l = l'$ or $l \parallel l'$

One way to think of rotations, translations, and reflections is to think of yourself as an observer situated at a particular point of the plane. Now suppose that you turn clockwise through a 90° angle. Then the appearance of the plane will be precisely the same as if the plane had been rotated about you through a 90° counterclockwise angle while you remained stationary. (Explain how Foucault's pendulum illustrates this concept.) Or, suppose that you move a certain distance to your right. Then from your new position the appearance of the plane will be the same as if the points of the plane had been translated that same distance to the left.

The diagram below suggests that if an observer properly changes his position in space, then the appearance of a plane is changed in exactly the same manner as if the plane were reflected in a line. Explain how the diagram shows this. *Hint:* To the observer at P the vertices A, B, C of $\triangle ABC$ seem to be in clockwise order.]

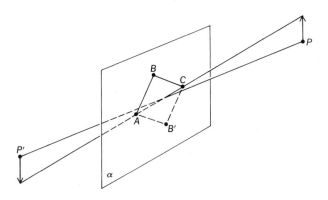

The point of view we have sketched strengthens intuitive understanding of transformations. The invariance of certain geometric properties under the rigid motions seems quite apparent. Intersecting lines are mapped upon lines which intersect at the same angle; parallels are mapped upon parallels.

One of the chief theorems of Euclidean geometry is that any transversal of two parallel lines forms congruent corresponding angles. For example, in the figure below, $l \parallel m$ and $\angle 1$ corresponds to $\angle 2$; hence $\angle 1 \cong \angle 2$.

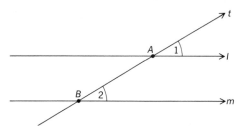

For a proof of this theorem simply apply the translation determined by the vector from A to B. The point A is mapped upon B, line t is mapped upon itself, and l upon m. "Clearly" $\angle 1$ is mapped upon $\angle 2$, and so $\angle 1 \cong \angle 2$. This "proof" deserves a little more attention. We know that the translation maps l upon a line l' through B such that $l' \parallel l$. (Why?) In order to be sure that $l' = m$, we need to know that m is the *only* line of the plane through B and parallel to l. This is a key assumption we must make. Euclid's fifth postulate is needed for our work with transformations.

We give another illustration of the use of transformations in proofs. Consider the theorem that base angles of an isosceles triangle are congruent.

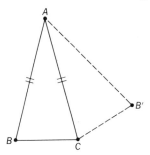

Explain how the proof that $\angle B \cong \angle C$ follows from a consideration of the composition of the two transformations below.

a) Reflect the plane in $\overleftrightarrow{AC}$.

b) Rotate ~~counter~~clockwise about A through $\angle BAC$.

The exercises below present a series of problems whose solutions utilize properties of transformations that we have mentioned. One property that we will use repeatedly is that if l is a line and A is a point not on l, then the 180° rotation about A maps l upon a line l' parallel to l. Can you prove that this is so?

Exercises 12–8

1. Use the 180° rotation H_M to prove $\angle 1 \cong \angle 2$, and $\angle 3 \cong \angle 4$. Hence, vertical angles are congruent. What do you prove by reflecting the plane in the line that bisects $\angle 1$? How do you know that the line that bisects $\angle 1$ also bisects $\angle 2$?

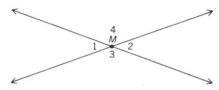

2. Use the transformation H_N to prove that the alternate interior angles, $\angle 1$ and $\angle 2$, are congruent if $l \parallel m$. How do you know that $H_N(l) = m$, and $H_N(m) = l$?

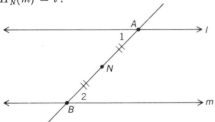

3. Ray $A\vec{M}$ bisects angle A of isosceles triangle ABC. Use the reflection in $\overset{\leftrightarrow}{AM}$ to prove that $\overset{\leftrightarrow}{AM} \perp BC$ and $\triangle AMB \cong \triangle AMC$. Note that it is not given that M is the midpoint of $\overline{BC}$. This comes out in the proof.

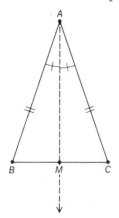

4. In $\triangle ABC$, M is the midpoint of $\overline{AB}$; and N, the midpoint of $\overline{AC}$. Show that the half-turn H_M maps $\angle ABC$ upon $\angle BAC'$ and that the half-turn H_N maps $\angle ACB$ upon $\angle CAB'$. Hence the sum of the angles of a triangle is a straight angle.

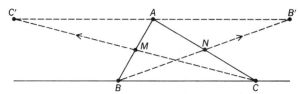

5. An exterior angle of a triangle is greater than either of the opposite interior angles.

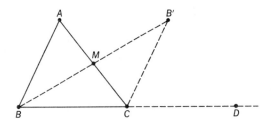

In $\triangle ABC$, let M be the midpoint of $\overline{AC}$. Show that the half-turn H_M maps $\angle BAC$ upon $\angle B'CA$ which is smaller than exterior angle, $\angle ACD$. How would you show that $\angle B < \angle ACD$?

6. In any parallelogram $ABCD$, opposite sides are congruent, opposite angles are congruent, each diagonal divides the parallelogram into congruent triangles, and the diagonals bisect each other.

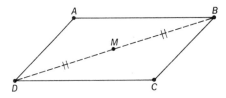

It is given that $\overleftrightarrow{AB} \parallel \overleftrightarrow{DC}$ and $\overleftrightarrow{DA} \parallel \overleftrightarrow{CB}$. Let M be the midpoint of $\overline{DB}$. The half-turn H_M maps each line l upon a line l' parallel to l. Complete the proof. [*Hint:* If you can show that H_M maps A upon C, you will know that M is the midpoint of diagonal $\overline{AC}$. Why?]

7. Show how to use *translations* to prove that opposite sides of a parallelogram are congruent. Note that the translation T with vector **v** maps A upon D.

But this translation has the same *direction* as **v**′. (Why?) So if T maps B upon a point of line $\overleftrightarrow{DC}$, this point must be C. Complete the proof.

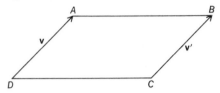

8. Show that if the diagonals of a quadrilateral bisect each other, the quadrilateral is a parallelogram.

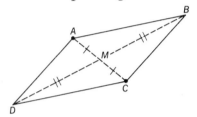

Consider H_M the rotation of 180° about the intersection of the diagonals.

9. Prove that if two sides of a quadrilateral are congruent and parallel, the quadrilateral is a parallelogram.

It is given that $\overline{AB} \cong \overline{DC}$ and $\overleftrightarrow{AB} \parallel \overleftrightarrow{DC}$. Rotate the plane 180° around M and note that $\overleftrightarrow{BA}$ is mapped upon a line through D, parallel to $\overleftrightarrow{BA}$.

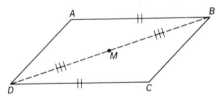

10. Prove that the line joining midpoints of opposite sides of a parallelogram is parallel to the other two sides.

Note that $\overleftrightarrow{MN}$ is obtained from $\overleftrightarrow{AB}$ by a translation.

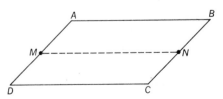

11. Prove that the line joining midpoints of two sides of a triangle is parallel to and half as long as the third side. Apply the half-turn H_M. Show that $C'ACB$ is a parallelogram. Use Exercise 9 to show that $N'NCB$ is a parallelogram, and complete the proof.

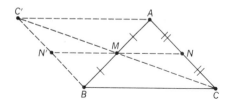

12. Show that the composition of two half-turns $H_N H_M$ is a translation in the direction from M to N and twice the distance from M to N. The figure suggests a proof. Explain.

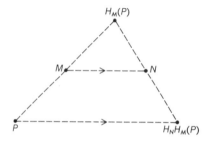

13. Show that the quadrilateral formed by joining the midpoints of the sides of *any* quadrilateral is a parallelogram. Note that

$$H_N H_M(A) = C, \qquad \text{and} \qquad H_O H_P(A) = C.$$

So the vector from M to N must have the same direction and length as the vector from P to O. (Why?)

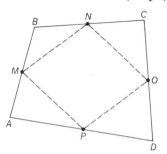

14. Use the definition that a circle is the set of all points in the plane at a fixed distance from its center.

a) Prove that the circle with center O is mapped upon itself by every rotation about O.

b) Prove that a circle is mapped upon itself by the reflection in any line passing through O.

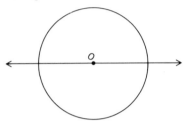

15. Assume that at each point T of a circle the circle has a *unique* tangent line t which intersects the circle in the single point T.

Prove that the radius $\overline{OT}$ is perpendicular to t. [*Hint:* Reflect the plane in $\overleftrightarrow{OT}$, and show that t must be mapped upon itself. Hence $\angle OTR$ is mapped upon $\angle OTS$.]

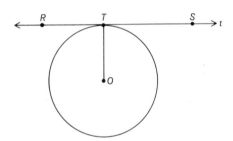

16. Let P be in the interior of $\angle B$. Construct a line segment with endpoints on the sides of $\angle B$, having P for midpoint. [*Hint:* Note that if P is the midpoint of $\overline{XY}$, then the half-turn H_P maps B upon B' so that $BXB'Y$ is a parallelogram. Describe the construction for points X and Y.]

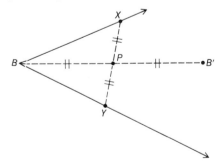

17. Let l and m be two intersecting lines, and P, Q, R three points none of which lies on l or m. Construct a path consisting of three line segments with the following properties.

a) The path begins at a point $L \in l$ and ends at $M \in m$.

b) The path passes in turn through P, Q, R.

c) Each of P, Q, R is the midpoint of the segment containing it.

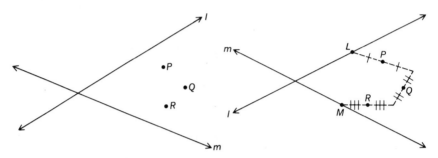

[*Hint:* Note that the composite, $H_R H_Q H_P$, of the three half-turns about P, Q, and R is itself a half-turn which maps L upon M and M upon L. (Why?) Hence, the midpoint of $\overline{LM}$ is the center of rotation for the half-turn $H_R H_Q H_P$. Find this center of rotation and use the result of Exercise 16 to complete the construction.]

18. Let l be any line and let A and A' be two points such that $M_l(A) = A'$. For any point B in the plane, show how to construct with straightedge alone the point

$$B' = M_l(B).$$

[*Hint:* If the lines $\overleftrightarrow{A'B}$ and $\overleftrightarrow{AB'}$ intersect, where is this point of intersection?]

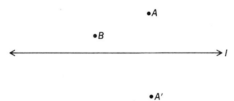

19. Let $\triangle ABC$ be any triangle, and let points P, Q, R be midpoints of the sides, as indicated. Verify that

a) $H_P H_Q H_R = H_A = H_R H_Q H_P$,

b) $H_Q H_R H_P = H_B = H_P H_R H_Q$,

c) $H_R H_P H_Q = H_C = H_Q H_P H_R$.

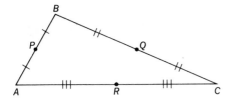

20. Prove the side-side-side congruence theorem for triangles by completing the argument begun below.

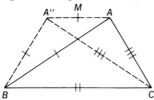 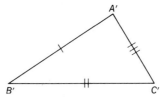

For triangles ABC and $A'B'C'$, let the sides be congruent in pairs as indicated. Move $\triangle A'B'C'$ so that $\overline{B'C'}$ coincides with $\overline{BC}$ and A' falls upon A'' on A's side of $\overleftrightarrow{BC}$. Then it must be the case that $A'' = A$. For if not, let M be the midpoint of $A''A$. Consider lines $\overleftrightarrow{BM}$ and $\overleftrightarrow{CM}$. Both lines are perpendicular to $\overline{A''A}$. (Why?)

13 Coordinate Geometry

13–1 Coordinate Systems for a Line

The axioms of geometry enable us to establish a 1–1 correspondence between the real numbers and the points of any line. We pick any two points A and B on the line and associate O with A and 1 with B.

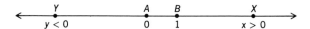

If X is any point on B's side of A, we associate with X the positive real number that is the length of $\overline{AX}$ in terms of unit segment $\overline{AB}$. If Y is on the ray opposite $\overleftrightarrow{AB}$, we associate with Y the negative real number that is the negative of the length of $\overline{AY}$. Each correspondence of this type is called a *coordinate system* for the line. Point A is the *origin*, and B is the *unit point*. Each associated number is the co-ordinate of its point. Often we shall identify points and their coordinates; for example, we may speak of "point 3" when we mean the "point whose coordinate is 3." We shall use small letters $a, b, c, \ldots$ to denote, respectively, the coordinates of points denoted by $A, B, C, \ldots$

Exercises 13–1

1. Three points A, B, C have, respectively, the coordinates 4, -7, -3. Which point is between the other two?

2. Points A, B, C have coordinates a, b, c, respectively, with $a < b$. Which of these three points is between the others if

   a) $b < c$? b) $c < a$? c) $c > a$ and $c < b$?

3. If Y is between X and Z, what inequalities hold for the coordinates x, y, and z?

4. Compute the length of $\overline{AB}$, given that

   a) $a = 4$, $b = 10$,   b) $a = -4$, $b = 10$,

   c) $a = 4$, $b = -10$,   d) $a = -4$, $b = -10$.

5. Verify that for every part of Exercise 4,

   $$m(\overline{AB}) = |b - a| = |a - b|.$$

   State and prove a general theorem.

6. M is given as the midpoint of $\overline{AB}$. What is m if

   a) $a = 4$, $b = 10$?   b) $a = 4$, $b = -10$?

   c) $a = -4$, $b = 10$?   d) $a = -4$, $b = -10$?

7. Verify that for every part of Exercise 6, $m = (a + b)/2$. State and prove a general theorem.

8. Let T_1 and T_2 be points of trisection for the segment $\overline{AB}$. Compute t_1 and t_2, given that

   a) $a = 6$, $b = 15$,   b) $a = 6$, $b = -15$,

   c) $a = -6$, $b = 15$,   d) $a = -6$, $b = -15$.

9. Verify that for every part of Exercise 9 the formulas

   $$t_1 = \frac{2a + b}{3}, \qquad t_2 = \frac{a + 2b}{3}$$

   furnish the coordinates of T_1 and T_2. Guess formulas for the coordinates of the 4 points that divide a line segment into 5 congruent parts, and check your formulas, using as an example the segment $\overline{AB}$ with $a = -3$, $b = 12$.

10. Let $a = -3$ and $b = 7$. Determine c so that

    a) B is the midpoint of $\overline{AC}$,

    b) C is the midpoint of $\overline{BA}$,

    c) A is the midpoint of $\overline{BC}$.

11. It is convenient to define the directed distance from one point on a line to another. This directed distance may be either positive or negative. For example, in the figure below, the directed distance from R to S is 5, and the directed distance from S to R is -5.

   a) In the figure above give the directed distances

   AD, DA, PC, CP, BE, EB.

b) With the notation we have agreed upon, show that

$$AD = d - a, \qquad DA = a - d, \qquad RC = c - r.$$

c) If points X and Y on a line have coordinates x and y, respectively, what is the formula for XY? For YX?

In the remainder of this chapter we shall use the symbol RS to denote the directed distance *from point R to point S* on a coordinate line. Note that with this notation we have

$$m(\overline{XY}) = |XY|,$$

that is, the length of line segment $\overline{XY}$ on a coordinate line is the *absolute value* of the directed distance from X to Y.

13–2 Coordinate Systems for a Plane

We get a coordinate system for a plane by choosing two perpendicular lines and establishing a coordinate system on each line. The point of intersection of the lines is taken as the origin for each coordinate system on these lines. Ordinarily the same unit of length is chosen for these coordinate systems. In the pictures that we draw of coordinate systems for a plane, one line is usually drawn horizontal and called the X-axis. The vertical line is then called the Y-axis.

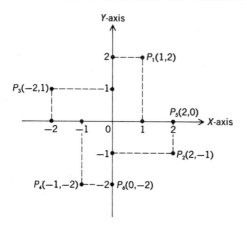

Now each point in the plane can be named by an ordered pair of real numbers. The figure shows points $P_1, \ldots, P_6$ named by $(1, 2), \ldots, (0, -2)$. We say that the x-coordinate (or first coordinate) of P_1 is 1 and the y-coordinate (or second coordinate) is 2. Explain how these coordinates are determined by drawing lines through P_1 parallel to the coordinate axes. Working with a coordinate system for a plane may be thought of as simultaneously working with coordinate systems on horizontal lines and on vertical lines. We illustrate this below.

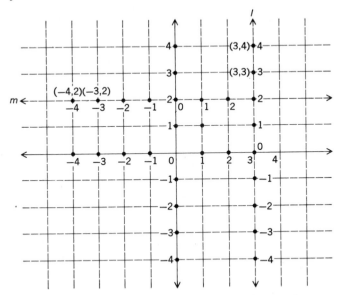

We can think of each vertical line (as l) and each horizontal line (as m) as possessing a coordinate system of its own matching the Y-axis and X-axis coordinate systems, respectively. Every point on line l above has first coordinate or X-coordinate 3, and every point on m has second coordinate or Y-coordinate 2. The equation of l is $X = 3$; and of m, $Y = 2$. Consider the problem of computing the midpoint of any segment in the plane.

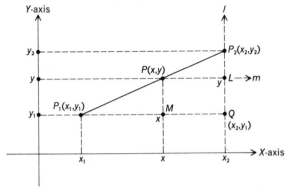

Let P_1, P_2 be any two points in the plane with coordinates (x_1, y_1) and (x_2, y_2), respectively. Then, as the diagram suggests, in order to find the coordinates (x, y)

of the midpoint P of $\overline{P_1P_2}$, we must find the x-coordinate of point M and the y-coordinate of point L. To compute x, we look at line m and ignore the y-coordinates of points on this line. M is the midpoint of $\overline{P_1Q}$; so

$$x = \frac{x_1 + x_2}{2}.$$

(Explain.) Similarly,

$$y = \frac{y_1 + y_2}{2}.$$

We get the midpoint of a segment in the plane by using twice the familiar midpoint formula for coordinate systems on a line.

If $\quad P_1 = (x_1, y_1), \; P_2 = (x_2, y_2),$

then

$$\left(\frac{x_1 + x_2}{2}, \frac{y_1 + y_2}{2}\right) \text{ is the midpoint of } \overline{P_1P_2}.$$

We get the formula for the distance between any two points in the plane by using the formula for the distance between two points on a line and then applying the Pythagorean theorem. If P_1 and P_2 are on a vertical or horizontal line, it is easy to compute this distance. (Explain.) If this is not the case, then the vertical line through P_1 and the horizontal line through P_2 intersect at a point Q. The coordinates of Q are (x_1, y_2). (Explain.) As the figure shows, the directed distances P_1Q and QP_2 are respectively $y_2 - y_1$ and $x_2 - x_1$. (Are these directed distances positive or negative in the figure?) Now the length of $\overline{P_1P_2}$ is given by

$$m(\overline{P_1P_2}) = \sqrt{(x_2 - x_1)^2 + (y_2 - y_1)^2}$$
$$= \sqrt{(x_1 - x_2)^2 + (y_1 - y_2)^2}.$$

Explain the last equality above.

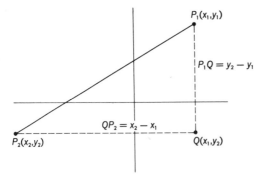

Referring to the same figure, we find that the *slope* of line $\overset{\leftrightarrow}{P_1P_2}$ is "rise over run" or

$$\frac{y_2 - y_1}{x_2 - x_1}.$$

Note that in our figure $y_2 - y_1$ is negative and $x_2 - x_1$ is negative, so the slope is a positive number. Show that if you think of traversing the line $\overset{\leftrightarrow}{P_1P_2}$ from P_2 to P_1, then it is natural to write the slope as

$$\frac{y_1 - y_2}{x_1 - x_2}.$$

Explain why

$$\frac{y_1 - y_2}{x_1 - x_2} = \frac{y_2 - y_1}{x_2 - x_1}.$$

It follows that in the formula for the slope m_l of a line l through the two points P_1, P_2,

$$m_l = \frac{y_2 - y_1}{x_2 - x_1}$$

and it does not matter which point we consider to be our first point and which the second.

Problem. Sketch a figure to show a line l with slope

$$\frac{y_2 - y_1}{x_2 - x_1}$$

determined by two points P_1 and P_2. If two other points P_1' and P_2' on the line are chosen with coordinates (x_1', y_1') and (x_2', y_2'), respectively, then show that

$$\frac{y_2' - y_1'}{x_2' - x_1'} = \frac{y_2 - y_1}{x_2 - x_1}.$$

Hence we get the same number for the slope of a line regardless of the points we choose on the line. [*Hint:* You will need to use properties of similar triangles in making your argument.]

Note that in the formula

$$m_l = \frac{y_2 - y_1}{x_2 - x_1}$$

for the slope of a line l, if line l is a vertical line, then the formula is not usable. (Why?) Vertical lines do not have slopes. However, a horizontal line has a slope. Explain how the formula is applicable to give the slope of a horizontal line.

Using a coordinate system, one can prove easily many geometric theorems. We illustrate this in the following examples. We assume the following theorems:

1. Two lines are parallel if and only if they have equal slopes or are both vertical lines.

2. Two lines are perpendicular if and only if their slopes are negative reciprocals of each other, or if one line is horizontal and the other vertical. For example, lines of slopes

$$\frac{2}{5} \quad \text{and} \quad \frac{-5}{2} \quad \text{are perpendicular,}$$

as are lines of slopes $\quad \dfrac{-8}{3} \quad$ and $\quad \dfrac{3}{8}.$

Theorem. The segments joining the midpoints of a quadrilateral in order, form a parallelogram.

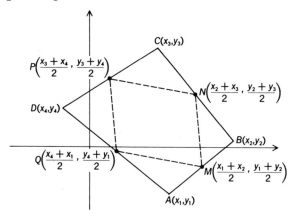

Proof. Let A, B, C, D be the vertices of a quadrilateral with coordinates as shown in the figure. Let M, N, P, Q be the midpoints of the sides as shown. The coordinates of these midpoints are indicated in the figure. (Explain.) Now we compute the slopes of $\overleftrightarrow{NM}$ and $\overleftrightarrow{PQ}$. (Explain this computation.)

$$m_{\overleftrightarrow{NM}} = \frac{\dfrac{y_2 + y_3}{2} - \dfrac{y_1 + y_2}{2}}{\dfrac{x_2 + x_3}{2} - \dfrac{x_1 + x_2}{2}} = \frac{y_3 - y_1}{x_3 - x_1}.$$

$$m_{\overleftrightarrow{PQ}} = \frac{\dfrac{y_3 - y_4}{2} - \dfrac{y_4 + y_1}{2}}{\dfrac{x_3 + x_4}{2} - \dfrac{x_4 + x_1}{2}} = \frac{y_3 - y_1}{x_3 - x_1}.$$

Since

$$m_{\overleftrightarrow{NM}} = m_{\overleftrightarrow{PQ}},$$

the lines are parallel. Similarly, we can show that $\overleftrightarrow{PN} \parallel \overleftrightarrow{QM}$. The quadrilateral $QMNP$ is a parallelogram because the opposite sides are parallel.

Theorem. The diagonals of a square are perpendicular.

Proof: We choose the coordinate axis as shown in the figure. Then the slope of one diagonal is 1 and the other is -1. (Explain.) Hence the diagonals are perpendicular. (Why?)

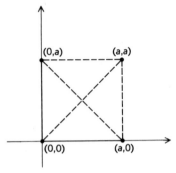

Theorem. The line segment from the vertex of the right angle of a right triangle to the midpoint of the hypotenuse is half as long as the hypotenuse.

Proof. The figure and the calculation below present the proof. (Explain.)

$$\sqrt{(a-0)^2 + (b-0)^2} = \sqrt{(a-2a)^2 + (b-0)^2}.$$

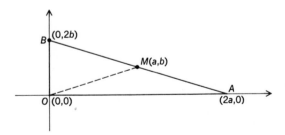

Exercises 13–2

1. Give the slopes of the lines through the following pairs of points.

  a) $(3, 5)$ and $(-1, 8)$ b) $(-2, 6)$ and $(1, 10)$

  c) $(7, 5)$ and $(-1, 11)$ d) $(3, -5)$ and $(5, -2)$

  e) $(3, 7)$ and $(5, 7)$ f) $(7, 3)$ and $(7, 5)$.

2. In Exercise 1 show that the lines of (a) and (c) are parallel; that those of (a) and (b) are perpendicular; that those of (a) and (d) are not parallel. What can you conclude about the lines of (b) and (c)? Of (c) and (d)? Of (e) and (f)?

3. Give the distance between the points of each of the following pairs.

   a) $(4, -2), (7, 2)$

   b) $(4, -2), (-7, 2)$

   c) $(4, -2), (7, -2)$

   d) $(4, -2), (4, 8)$

4. Prove that all the points $(6, 3), (6, -5), (3, -6), (8, -1), (-1, 2), (-1, -4)$ lie on a circle with center $(3, -1)$. What is the radius of this circle? Plot all these points on graph paper.

5. What can you say about the slope of a line that is very nearly vertical? Very nearly horizontal?

6. Let A and B have coordinates $(3, 2)$ and $(5, 8)$. Find the coordinates of C so that

   a) B is the midpoint of $\overline{AC}$.

   b) C is the midpoint of $\overline{BA}$.

   c) A is the midpoint of $\overline{BC}$.

7. A line with slope $\frac{3}{4}$ contains the origin. Give coordinates of two other points on the line.

8. A line with slope $-\frac{1}{2}$ contains the point $(1, 5)$. Give coordinates of two other points on the line.

9. Using graph paper, sketch the lines through the origin with slopes: $0, \frac{1}{2}, 1, 2, 4, 8, -\frac{1}{2}, -1, -2, -4, -8$.

10. Determine whether or not the three points are collinear. [*Hint:* Use ideas of slope.]

    a) $(-1, -12), (1, -4),$ and $(3, 5)$

    b) $(1, 3), (-1, 7),$ and $(3, -1)$

    c) $(6, -3), (2, 5),$ and $(-1, 16)$

11. Show that the triangles with the given vertices are isosceles.

    a) $A(6, 1), B(2, -4), C(-2, 1)$

    b) $A(-3, -3), B(3, 3), C(-4, 4)$

12. Show that the triangles with the given vertices are right triangles.

    a) $A(1, 4), B(10, 6), C(2, 2)$

    b) $A(0, 4), B(-3, -3), C(2, -1)$

13. Show how to use the distance formula, and decide whether the points lie on a straight line.

    a) $(3, 2), (0, 0), (9, 6)$

    b) $(-4, 0), (0, 2), (9, 7)$

    c) $(2, 1), (-1, 2), (5, 0)$

14. If the point $(x, 3)$ is equidistant from $(3, -2)$ and $(7, 4)$, what is x?

15. Find the point on the y-axis equidistant from $(-5, -2)$ and $(3, 2)$.

16. Use the midpoint formula to prove that the diagonals of a parallelogram bisect each other.

    [*Hint:* Choose the X- and Y-axes as shown. What are the coordinates of the 4th vertex?]

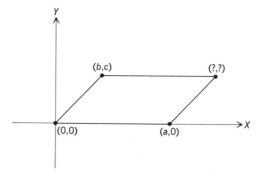

17. Prove that the line segment joining the midpoints of two sides of a triangle is parallel to the third side and half as long.

18. Prove that the sum of the squares on the sides of a parallelogram is equal to the sum of the squares on the diagonals.

19. Prove that if $(0, 0), (a, 0),$ and (b, c) are vertices of an isosceles triangle as shown in the figure, then $a = 2b$. Now prove that the line from the vertex (b, c) to the midpoint of the base is perpendicular to the base.

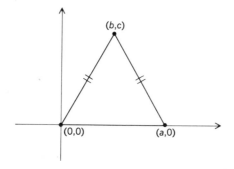

13–3 Linear Equations

In plane coordinate geometry, geometric figures correspond to equations in two variables. The fundamental problems of analytic geometry may be categorized as follows:

1. Starting with an equation, describe the corresponding geometric figure.

2. Starting with a geometric figure, find the associated equation.

The graph of an equation is the set of points whose coordinates satisfy the equation. For example, the points with coordinates

$$(2, 3),\ (1, 0),\ (\tfrac{3}{2}, \tfrac{3}{2}),\ (0, -3) \qquad \text{and} \qquad (\sqrt{2}, -3 + 3\sqrt{2})$$

are on the graph of the equation $3X - Y - 3 = 0$; and the points with coordinates

$$(0, 0),\ (1, 1),\ (-2, 4),\ (-\pi, \pi^2) \qquad \text{and} \qquad (\sqrt{2}, 2)$$

are on the graph of $Y = X^2$.

A linear equation in two variables is an equation of the form

$$ax + by + c = 0,$$

where a, b, and c are real numbers. In the next set of exercises, we study graphs of these equations.

Exercises 13–3

1. Complete the tables and graph the points.

a) $y = x + 2$

x	0	1	2	-2	
y					8

b) $y = x + 3$

x	0	1	2		
y				1	0

c) $y = x + 5$

x	1	2			-2
y			9	8	

d) $y = 2x$

x	0	1			
y			4	-8	

e) $y = 3x + 2$

f) $y = 5x + 2$

g) $y = 2x + 3$

2. What appears to be true for graphs of linear equations?

3. Some of the points on the graph of a line are given in (a), (b), (c), (d), and (e). Locate 2 or 3 additional points on each graph. Write a linear equation of the form $y = mx + d$ for each graph, and verify that at least 3 points of the line satisfy your equation.

[*Hint:* Try $y = 2x + d$ (for a). Try $y = 4x + d$ (for e).]

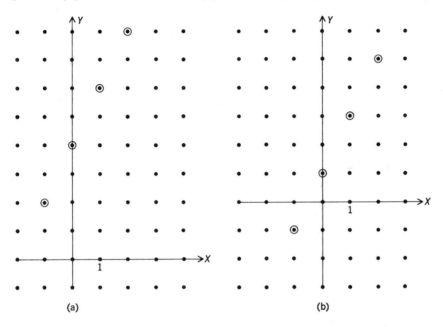

(a)

(b)

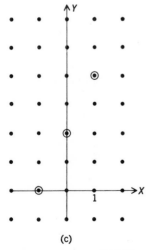

(c)

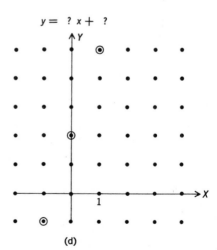

$y =$? $x +$?

(d)

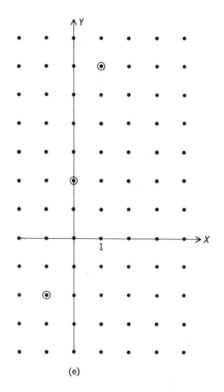

(e)

4. In Exercise 3, read off the slope of each line from its graph. Graph the four lines below and explain how to determine the slope of each line from its equation.

    a) $y = 2x - 3$ b) $y = 3x + 1$

    c) $y = -2x + 4$ d) $y = -3x - 1$

5. Show that the equation $y = 5$ has for its graph a horizontal line. Describe the graph of

    a) $x = -3$ b) $y = 0$ c) $x = 0$

6. When a linear equation can be written in the form $y = mx + d$, what is the geometric significance of the number m? Of d? Assuming that $b \neq 0$, write the equation $ax + by + c = 0$ in the above form. What is the slope of this line? At what point does it cross the y-axis?

7. What type of graph is associated with the equation $ax + by + c = 0$ if

    a) $a = 0, b \neq 0$ b) $b = 0, a \neq 0$

    c) $a = 0, b = 0, c \neq 0$ d) $a = 0, b = 0, c = 0$

8. If linear equations in two unknowns are to have line graphs, what restriction must be placed upon the coefficients of the equation $ax + by + c = 0$?

9. Show that

a) If (a, b) is on the graph of $2x - 3y + 5 = 0$, then so also is the point $(a + 3, b + 2)$.

b) If (a, b) is on the graph of $2x + 3y + 7 = 0$, then so also is $(a - 3, b + 2)$.

c) If (r, s) is on the line $ax + 3y + c = 0$, then so also is $(r - 3, s + a)$.

10. Prove that the graph of the equation $2x + 3y + 5 = 0$ of Exercise 9(a) is a straight line by arguing that

a) Every point on its graph lies on the line through $(-1, 1)$ with slope $\frac{2}{3}$, and conversely,

b) Every point on the line through $(-1, 1)$ with slope $\frac{2}{3}$ lies on the graph of the equation.

The last set of exercises indicates that straight lines have equations of the form $ax + by + c = 0$, where not both a and b are zero, and that conversely, any such equation has a straight line graph. This conclusion illustrates the comment made earlier concerning the two basic problems of analytic geometry. (Explain.) We shall not give a formal proof of this result. If you worked Exercise 10 above, you proved a special case of this fundamental theorem. We will use this result in problem solving. For example, we can determine an equation for the line through the points $(2, 3)$ and $(7, 1)$ as shown below:

1. We assume that an equation of the line has the form

$$ax + by + c = 0.$$

2. Then $2a + 3b + c = 0$, and $7a + b + c = 0$. (Why?)

3. Subtracting the first equation from the second, we get

$$5a - 2b = 0. \quad \text{(Explain.)}$$

4. If we take $a = 2$ and $b = 5$, our equation becomes

$$2x + 5y + c = 0.$$

5. Substituting 2 for x and 3 for y, we see we must choose $c = -19$ in order for the point $(2, 3)$ to satisfy the equation. Hence, a possible equation is $2x + 5y - 19 = 0$. Now we verify that $(7, 1)$ is on this line.

Problem. In the above example choose $a = 4$ and $b = 10$, and compare the resulting equation with $2x + 5y - 19 = 0$. How is it that one line can have many different equations?

It is convenient to know that if $ax + by + c = 0$ is not a vertical line, its slope is given by $-a/b$. Explain how the calculation below is a proof of this fact.

Let (x_1, y_1) and (x_2, y_2) be any two points on the line $ax + by + c = 0$. Then

1. $ax_1 + by_1 + c = 0$ and $ax_2 + by_2 + c = 0$. (Why?)
2. Hence

    $a(x_2 - x_1) + b(y_2 - y_1) = 0$. (Why?)

3. And

    $$\frac{y_2 - y_1}{x_2 - x_1} = \frac{-a}{b}. \quad (\text{Why?})$$

    [Note that $x_2 - x_1 \neq 0$. (Why?) Why is $b \neq 0$?]
4. Hence the slope of the line is

    $$\frac{-a}{b}. \quad (\text{Why?})$$

With this knowledge we can choose the coefficients for x and y if we know the slope of a line. For example, if the slope is $-3/5$ we can write the equation as

$$3x + 5y + c = 0.$$

If we know also that the line lies on the point $(0, 1)$, then $c = -5$ (Why?), and an equation is

$$3x + 5y - 5 = 0.$$

Exercises 13–4

1. Show that each of the lines $3x - 4y + 2 = 0$ and $-3x + 4y + 7 = 0$ has slope $\frac{3}{4}$.

2. Write equations of three lines each having slope
    a) $7/5$ b) $-4/9$

3. Determine an equation of a line having slope $-5/3$ and lying on the point $(1, -1)$.

4. The line through the points $(4, 3)$ and $(9, 5)$ has slope $2/5$. (Why?) And hence has an equation,

    $$2x - 5y + c = 0.$$

    Choose c so that $(4, 3)$ satisfies this equation, and show that $(9, 5)$ also satisfies the equation.

5. Find an equation of the line on

    a) $(-3, 1)$ and $(2, 5)$ b) $(0, 4)$ and $(-5, 0)$

    c) $(4, 3)$ and $(7, 3)$ d) $(3, 4)$ and $(3, 7)$

6. Find an equation of a line parallel to $3x - 4y + 7 = 0$ and passing through the point $(3, 2)$.

7. Find an equation of the line perpendicular to the line $2x - 5y + 11 = 0$ and on the point $(-2, 3)$.

8. Find the point of intersection for each pair of lines below.

    a) $2x + 3y = 7$ b) $5x - 3y = 13$

        $3y + 4x = -2$ $3x + 4y = 2$

    c) $y = 3x - 2$ d) $3x + 2y = 7$

        $y = 2x + 5$ $2x + 3y = 13$

9. Explain why the lines of each pair below are parallel.

    a) $2x + 3y = 7$ b) $y = 5x - 4$

        $2x + 3y = 10$ $5y - x = 7$

10. Compare the lines $3x - 5y + 6 = 0$ and $9x - 15y + 18 = 0$.

11. Find the perpendicular distance from the point $(5, 2)$ to the line

    $3x - 4y + 18 = 0$

    by writing an equation of the line through $(5, 2)$ perpendicular to

    $3x - 4y + 13 = 0$,

    finding the point of intersection of these lines, and computing the distance from $(5, 2)$ to this point.

12. What is the shortest distance

    a) From the origin to the line $3x + 4y = 24$,

    b) Between the lines $5x - 12y - 24 = 0$ and $5x - 12y + 2 = 0$?

13. A triangle has vertices $(2, 3)$, $(-1, 1)$, and $(7, -2)$. Find analytically the length of the altitude from the vertex $(2, 3)$ to the opposite side, the length of this opposite side, and the area of the triangle. Check the result by sketching the triangle on graph paper and computing the area by "counting" squares.

14. The points satisfying the equation

    $2x + 3y + 6 = 0$

    lie on a line. Do the points satisfying

    a) $2x + 3y + 6 > 0$ lie above or below this line?

    b) $2x + 3y + 6 < 0$ lie above or below this line?

15. Replace the line of Exercise 14 by $2x - 3y + 6 = 0$, and answer questions corresponding to (a) and (b).

16. Sketch the lines

$$3x + 4y - 18 = 0 \qquad \text{and} \qquad 5x + 2y - 16 = 0,$$

and shade in the following regions of the plane.

a) $3x + 4y - 18 > 0$ and $5x + 2y - 16 < 0$

b) $3x + 4y - 18 < 0$ and $5x + 2y - 16 > 0$

c) $3x + 4y - 18 > 0$ and $5x + 2y - 16 > 0$

d) $3x + 4y - 18 < 0$ and $5x + 2y - 16 < 0$

13-4 Transformations

We can describe transformations by equations. For example, on the line below if we denote by T_4 the translation that "moves" each point 4 units to the right, then for every point x

$$T_4(x) = x + 4$$

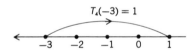

Verify that the reflection M_2 in the point 2 may be described by the equation

$$M_2(x) = -x + 4.$$

The composite transformation $M_2 T_4$ (first translate and then reflect) has for its equation,

$$M_2 T_4(x) = M_2(x + 4) = -(x + 4) + 4 = -x.$$

Note that $M_2 T_4$ is just the reflection in the origin. (Check this result geometrically.)

Problem. Compute $T_4 M_2(x)$. This is a reflection in what point?

Problem. Explain both algebraically and geometrically why $M_2 = T_2 M_0 T_{-2}$ and, in general, $M_a = T_a M_0 T_{-a}$. That is, to reflect the line in any point a, one can first apply the translation that moves the point a to the origin, then reflect in the origin, and then apply the translation that moves the origin back to a.

In the plane a translation may change both coordinates of a point. Just as T_a denotes the translation of a line which moves the origin to point a, so we shall

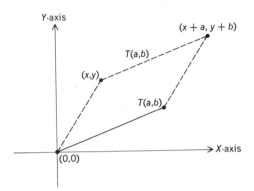

use $T_{(a,b)}$ to denote the translation of the plane that sends the origin to (a, b). As the diagram shows, the translation $T_{(a,b)}$ which sends the origin to the point (a, b) sends each point (x, y) to the point $(x + a, y + b)$. (Explain.) Hence the "equation" of this translation is

$$T_{(a,b)}(x, y) = (x + a, y + b).$$

Note that the *inverse* of $T_{(a,b)}$ is $T_{(-a, -b)}$. We sometimes describe a transformation by two equations. For example, the equations for $T_{(a,b)}$: $(x, y) \rightarrow (x', y')$ are

$$x' = x + a, \qquad y' = y + b.$$

Exercises 13–5

1. Using the facts that on a line the reflection in the origin M_0 has the equation $M_0(x) = -x$, and that for the reflection in a, $M_a = T_a M_0 T_{-a}$, show that $M_a(x) = -x + 2a$. Verify this result geometrically.

2. Verify that for transformations on a line,

   a) $T_a T_b = T_b T_a = T_{a+b}$,   b) $M_a M_b = T_{2(a-b)}$,

   c) $M_b M_a = T_{2(b-a)}$,   d) $T_{2b} M_a = M_{a+b}$,

   e) $M_a T_{2b} = M_{a-b}$.

3. Use the following technique to derive the formula for the reflection M_a in the point a of a line l.

   a) If $M_a(x) = x'$, then

   $$a = \frac{x + x'}{2}. \quad \text{(Why?)}$$

   b) Hence $x' = -x + 2a$. (Explain.)

4. Write the equation for the translation which maps

   a) $(0, 0)$ onto $(7, 3)$,   b) $(2, -6)$ onto $(3, -4)$,   c) (a, b) onto (c, d).

5. What is the relationship between the following pairs of translations?

   a) $T_{(3,5)}$ and $T_{(-5,3)}$ b) $T_{(-3,2)}$ and $T_{(-2,-3)}$

   c) $T_{(4,7)}$ and $T_{(-7,4)}$ d) $T_{(4,7)}$ and $T_{(-4,-7)}$

   e) $T_{(a,b)}$ and $T_{(-b,a)}$ f) $T_{(a,b)}$ and $T_{(-a,-b)}$

6. Show that

$$T_{(a,b)}T_{(c,d)} = T_{(a+c,b+d)}, \quad \text{and} \quad T_{(a,b)}T_{(c,d)} = T_{(c,d)}T_{(a,b)}.$$

7. Show that if $P = (x_1, y_1)$ and $Q = (x_2, y_2)$ are any two points in the plane, and $T_{(a,b)}$ is any translation, then if we set $T_{(a,b)}(P) = P'$, and $T_{(a,b)}(Q) = Q'$,

   a) $m(\overline{PQ}) = m(\overline{P'Q'})$ (i.e., translations preserve distance).

   b) Slope of $\overleftrightarrow{PQ}$ = slope of $\overleftrightarrow{P'Q'}$ (i.e., translations preserve slopes and hence send each line l to a line l' such that $l \parallel l'$).

8. Explain how translations can be used to show that plane figures are congruent.

9. Use a translation to show that the triangle with vertices $(4, 5)$, $(-1, 0)$, and $(3, -6)$ is congruent to the triangle with vertices $(7, 3)$, $(2, -2)$, and $(6, -8)$.

10. Is the triangle with vertices $(1, 1)$, $(6, 5)$, and $(-3, 4)$ congruent to the triangle with vertices $(0, 4)$, $(-4, 7)$, and $(5, 8)$?

11. Show that the figures in each case are congruent by giving the equation of a translation that maps one figure upon the other. Can you use more than one translation to show that the figures are congruent?

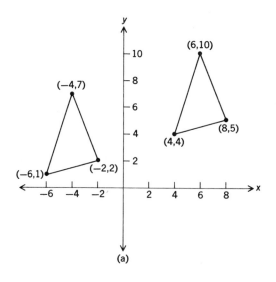

(a)

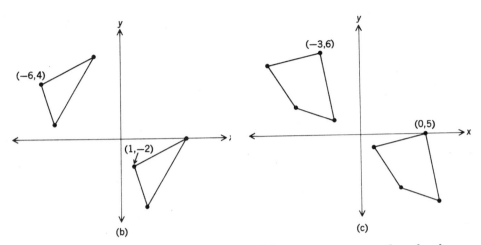

(b)   (c)

12. Assume that for two triangles it is impossible to map one upon the other by any translation. Does it follow that the triangles are not congruent? Explain your answer.

13. Show that there is a unique translation which maps (a, b) onto (c, d). How is the translation that maps (c, d) onto (a, b) related to this one?

14. Explain how a translation can be used to determine whether or not a quadrilateral is a parallelogram.

15. Determine whether or not the given points are vertices of a parallelogram.

   a) $\{(-5, -2), (-2, 3), (3, -2), (6, 3)\}$
   b) $\{(1, -7), (7, -6), (8, 7), (2, 6)\}$
   c) $\{(0, 1), (5, 2), (3, -5), (-3, -6)\}$

16. Find by translations a fourth point which determines a parallelogram with the points $(2, 1)$, $(3, 4)$, and $(8, 5)$. How many points have this property? Give the translation associated with each point.

Reflections in lines are more difficult to describe by equations than are translations. If P' is the image of P under the reflection in line l, then l is the perpendicular bisector of $\overline{PP'}$. The line l is called the *axis of the reflection*, and we denote the reflection by

$$M_l \qquad \text{or by} \qquad M_{ax+by+c=0},$$

if $ax + by + c = 0$ is the equation of l. Formulas for line reflections in vertical and horizontal lines are simple.

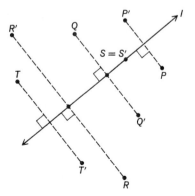

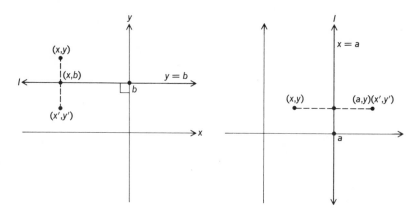

Consider the horizontal line l with equation $y = b$. Denote the image of (x, y) under $M_{y=b}$ by (x', y'). Clearly,

$$x = x' \qquad \text{and} \qquad b = \frac{y + y'}{2}.$$

(Explain.)

Show that the equations for the reflection $M_{y=b}$ are

$$\begin{cases} x' = x \\ y' = -y + 2b. \end{cases}$$

We leave it to the reader to find the expressions for a reflection in a vertical line $x = a$.

The reflection $M_{y=x}$ has a simple algebraic expression. Let

$$M_{y=x}(P) = P',$$

where P and P' have coordinates (x, y) and (x', y'), respectively, as in the figure.

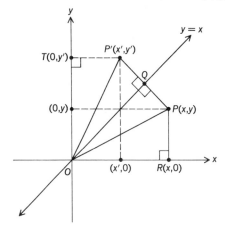

Now $\triangle PQO \cong \triangle P'QO$, so $\overline{OP} \cong \overline{OP'}$ and $\angle POR \cong \angle P'OT$.   (Explain.)   It follows that $\triangle POR \cong \triangle P'OT$, and hence $\overline{PR} \cong \overline{P'T}$ and $\overline{OR} \cong \overline{OT}$.   Thus we can conclude (why?) that

$$\begin{cases} x' = y \\ y' = x. \end{cases}$$

We can also write this as $M_{y=x}(x, y) = (y, x)$.

If l is any line with equation $ax + by + c = 0$, a tedious calculation produces the formula

$$M_l(x, y) = (x'y'),$$

where

$$x' = \frac{(b^2 - a^2)x - 2aby - 2ac}{a^2 + b^2}$$

and

$$y' = \frac{(a^2 - b^2)y - 2abx - 2bc}{a^2 + b^2}.$$

Problem. Verify for the formulas above that they agree with our earlier formulas for reflections in vertical lines, horizontal lines, and the line $y = x$.

Exercises 13–6

1. Find the images of the following points under reflection in the line $y = 2$.

   a) $(2, 0)$ b) $(-1, -1)$ c) $(6, -2)$,   d) $(4, 2)$

2. Find images of the points of Exercise 1 under reflection in

   a) $y = -2$ b) the y-axis   c) the x-axis

   d) $x = -3$ e) $y = x$ f) $y = -x$

   g) $3x + 2y - 1 = 0$

3. Complete the following: Under a line reflection,

   a) A line parallel to the axis of reflection is mapped into ...

   b) A line perpendicular to the axis of reflection is mapped into ...

   c) A line intersecting the axis of reflection at point P and forming an angle of 30° with the axis is mapped into ...

   d) Each point that is mapped upon itself lies on ...

4. Find the equations for the reflection in

   a) $x = a$,   b) $y = -x$,   c) the x-axis,   d) the y-axis.

5. Show that $\triangle ABC \cong \triangle A'B'C'$ by a line reflection (i.e., find the line reflection which maps $\triangle ABC$ onto $\triangle A'B'C'$).

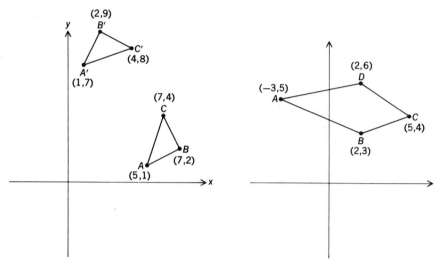

6. Find the vertices of a quadrilateral congruent to $ABCD$ by a line reflection in (a) the y-axis, (b) the x-axis, (c) the line $y = -x$.

7. Show that $\triangle ABC \cong \triangle A'B'C'$ by reflecting in the y-axis and then reflecting in the x-axis. What single line reflection can be used to obtain the above results?   *Point*

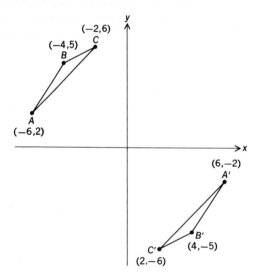

In Chapter 12 we noted that the composite of two reflections in parallel lines is a translation in a direction perpendicular to the lines and twice the distance between them. It was easy to deduce this result geometrically. We can use our equations for reflections to illustrate this fact for vertical lines. For example, if we reflect first in the vertical line $x = 3$ and then in the line $x = 5$, our equations are

$$M_{x=3}(x, y) = (-x + 6, y), \qquad M_{x=5}(x, y) = (-x + 10, y).$$

Consequently,

$$M_{x=5}M_{x=3}(x, y) = M_{x=5}(-x + 6, y) = (-(-x + 6) + 10, y) = (x + 4, y).$$

Explain the calculation above which shows that $M_{x=5}M_{x=3}$ is a translation, and show that this "analytic" result agrees with the "geometric" conclusions of the last chapter. Compute similarly $M_{x=3}M_{x=5}$, and interpret as a translation. How difficult would it be to establish this result analytically for parallel lines that are not horizontal or vertical?

We also noted in the last chapter that a reflection in a point may be considered either as a $180°$ rotation about the point or as the composite of two reflections in perpendicular lines through the point. (See the figure.)

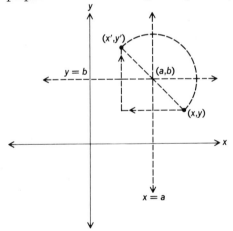

We can easily get the equations for the reflection in a point (a, b). Note that if (x', y') is the image of (x, y), then (a, b) is the midpoint of the segment having endpoints (x, y) and (x', y'). Hence

$$a = \frac{x + x'}{2}, \qquad b = \frac{y + y'}{2},$$

and

$$x' = -x + 2a, \qquad y' = -y + 2b.$$

Since $M_{x=a}(x, y) = (-x + 2a, y)$ and $M_{y=b}(x, y) = (x, -y + 2b)$, it follows that

$$M_{x=a}M_{y=b} = M_{y=b}M_{x=a} = \text{the reflection in } (a, b).$$

The reader should check the above calculations. This development illustrates how geometric transformations can be analyzed algebraically. A careful and complete treatment of these topics would involve us in algebraic calculations of considerable complexity.

Exercises 13–7

1. Verify that

$$M_{x=a}M_{y=b}(x, y) = M_{y=b}M_{x=a}(x, y) = (-x + 2a, -y + 2b).$$

2. Show that a point reflection preserves distances.

3. Show by a point reflection in $(-2, -2)$ that a triangle with vertices $(-7, 2)$, $(-4, 1)$, and $(1, 6)$ is congruent to a triangle with vertices $(3, -6)$, $(0, -5)$, and $(-5, -10)$.

4. Find the point reflection which maps $\triangle ABC$ onto $\triangle DEF$. Write the correct statement of congruence for these triangles.

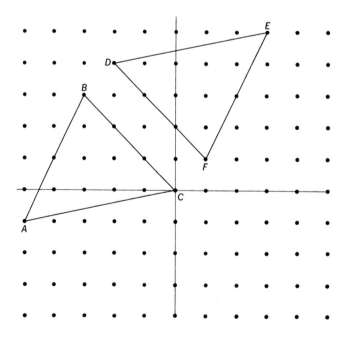

5. Show that for a 90° counterclockwise rotation about the origin, each point (x, y) is mapped upon the point $(-y, x)$. That is, $R_{90°}(x, y) = (-y, x)$. Now explain the algebraic argument that

$$R_{180°}(x, y) = R_{90°}R_{90°}(x, y) = R_{90°}(-y, x) = (-x, -y),$$

and verify this result geometrically.

6. In the chapter on geometric transformations we noted that the composite of two reflections in lines that form an angle of 45° is a 90° rotation about the point of intersection of the lines. The line $y = x$ forms a 45° angle with both the x-axis and the y-axis. Hence each of

$$M_{y=x}M_{y=0}, \quad M_{y=x}M_{x=0}, \quad M_{y=0}M_{y=x}, \quad M_{x=0}M_{y=x}$$

is a 90° rotation, clockwise or counterclockwise, about the origin. Verify algebraically that this is so, using the facts that

$$M_{y=x}(x, y) = (y, x), \qquad M_{y=0}(x, y) = (x, -y), \qquad M_{x=0}(x, y) = (-x, y),$$
$$R_{90°}(x, y) = (-y, x), \qquad R_{-90°}(x, y) = (y, -x).$$

7. Verify that a counterclockwise rotation of 90° about the point $(8, 5)$ has the equations

$$x' = 13 - y \qquad \text{and} \qquad y' = x - 3.$$

Use this result to verify that the pictured triangles are congruent.

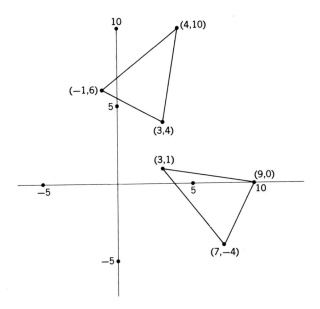

8. Compute the equations for a clockwise rotation of 90° about the point (2, 3) by use of reflections in the lines $y = 3$ and $y - 3 = x - 2$. (Use the equations given earlier for the reflection in any line.) Which reflection must be performed first? Check a few points.

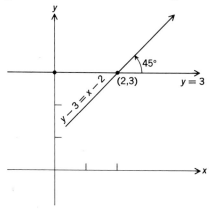

9. On graph paper draw a triangle ABC. Consider any point P and reflect it successively in A, in B, in C, in A, in B, in C.

a) What is the final image of P under this chain of six point reflections?

b) Write a theorem.

c) Prove your theorem using equations for point reflections. [*Hint:* Label your triangle $A(a_1, b_1)$, $B(a_2, b_2)$, $C(a_3, c_3)$, and let P have coordinates (x, y).]

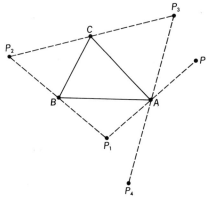

10. Transformations of the plane can be described by matrices. For example, the row-by-column technique for multiplying matrices enables us to make the following computation.

$$(x, y)\begin{pmatrix} 0 & 1 \\ 1 & 0 \end{pmatrix} = (y, x), \qquad (x, y)\begin{pmatrix} 0 & 1 \\ -1 & 0 \end{pmatrix} = (-y, x).$$

Explain the previous computation and show that the matrix

$$\begin{pmatrix} 0 & 1 \\ 1 & 0 \end{pmatrix}$$

represents the reflection in the line $y = x$. What transformation is represented by the matrix

$$\begin{pmatrix} 0 & 1 \\ -1 & 0 \end{pmatrix} ?$$

11. Find a matrix

$$\begin{pmatrix} a & b \\ c & d \end{pmatrix}$$

which represents the reflection in the x-axis. That is, we must have

$$(x, y)\begin{pmatrix} a & b \\ c & d \end{pmatrix} = (x, -y).$$

What matrix represents reflection in the y-axis?

12. Find images of the points $(2, 3)$, $(0, 4)$, $(5, -5)$ under the transformation represented by the matrix

$$\begin{pmatrix} 0 & -1 \\ -1 & 0 \end{pmatrix}.$$

This is a reflection in what line?

13. Show that the matrix for the 90° clockwise rotation about the origin is

$$A = \begin{pmatrix} 0 & -1 \\ 1 & 0 \end{pmatrix}.$$

Compute A^2 and show that this is the matrix for the 180° rotation. What is A^3?

14. The matrix

$$\begin{pmatrix} \dfrac{b^2 - a^2}{a^2 + b^2} & \dfrac{-2ab}{a^2 + b^2} \\[2ex] \dfrac{-2ab}{a^2 + b^2} & \dfrac{a^2 - b^2}{a^2 + b^2} \end{pmatrix}$$

represents the reflection in the line $ax + by = 0$, that is, the line through the origin with slope $-a/b$. Verify that for the line $y = 2x$ this matrix is

$$\begin{pmatrix} -3/5 & 4/5 \\ 4/5 & 3/5 \end{pmatrix}$$

and compute the effect of transforming each of the points $(0, 0)$, $(1, 2)$, $(3, 6)$, $(3, -4)$, $(4, 3)$, $(-5, 0)$, $(0, 5)$. Verify that the images of these points under the matrix multiplication are indeed their reflections in the line $y = 2x$.

15. Write out the eight matrices that represent the symmetries of the square pictured, and satisfy yourself by matrix multiplication that these matrices form a group of eight elements.

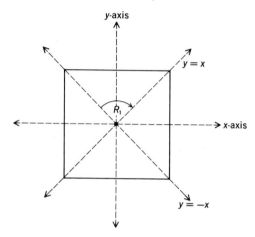

13–5 Graphs of Quadratic Equations

The graphs of first-degree (linear) equations are straight lines. The graphs of second-degree (quadratic) equations are called *conic sections*. This terminology is used because each of these graphs can be visualized as the intersection of a plane and a *complete* cone. A "complete" cone is illustrated in the accompanying figure: two infinite conical surfaces tip to tip.

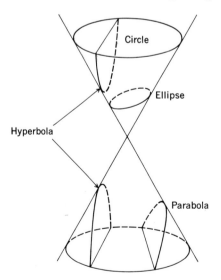

If the cone is circular, the various conic sections, circle, ellipse, parabola, and hyperbola, are sections of the cone as shown above. (Explain.)

An analysis of the properties of these second-degree curves is the chief task in an elementary analytic geometry course. We shall do little more than note the simplest equations for conics.

For a circle with center at the origin and radius r, the equation is $x^2 + y^2 = r^2$.

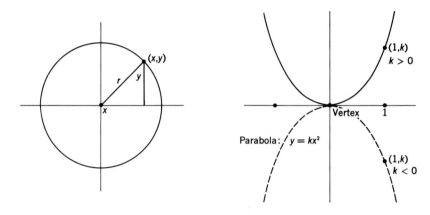

Circle: $x^2 + y^2 = r^2$ Parabola: $y = kx^2$

For a parabola with vertex at the origin and open "up," the equation is $y = kx^2$ with k *positive*. If $k < 0$ the parabola is open "down." Parabolas $x = ky^2$ are open to the right or left.

An ellipse with center at the origin and *foci* on the x-axis has an equation of the form

$$\frac{x^2}{a^2} + \frac{y^2}{b^2} = 1,$$

where $a > b$. If $a < b$ the foci are on the y-axis.

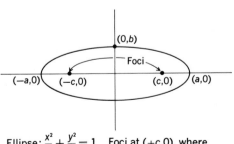

Ellipse: $\frac{x^2}{a^2} + \frac{y^2}{b^2} = 1$ Foci at $(\pm c, 0)$, where
$$a^2 = b^2 + c^2$$

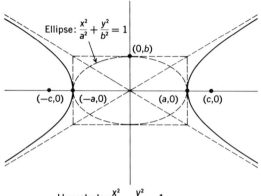

Ellipse: $\dfrac{x^2}{a^2} + \dfrac{y^2}{b^2} = 1$

(0,b)

(−c,0) (−a,0)   (a,0)   (c,0)

Hyperbola: $\dfrac{x^2}{a^2} - \dfrac{y^2}{b^2} = 1$

Foci at $(\pm c,0)$, where $c^2 = a^2 + b^2$

The graph of the curve

$$\frac{x^2}{a^2} - \frac{y^2}{b^2} = 1$$

is a hyperbola, center at the origin, and foci at $(\pm c, 0)$, where $c^2 = a^2 + b^2$. Note the relationship between the graphs of this hyperbola and the ellipse.

For the hyperbola

$$\frac{y^2}{a^2} - \frac{x^2}{b^2} = 1$$

the foci lie on the y-axis. Each hyperbola has two *asymptotes*, lines which intersect at the center and enclose the two branches of the curve.

Exercises 13–8

1. Sketch each circle giving center and radius.

   a) $x^2 + y^2 = 25$ b) $(x - 2)^2 + (y - 3)^2 = 9$

2. Sketch each parabola.

   a) $y = x^2$ b) $y = -x^2$ c) $x = 2y^2$ d) $x = -2y^2$

3. Sketch each ellipse and locate its foci.

   a) $\dfrac{x^2}{25} + \dfrac{y^2}{9} = 1$ b) $\dfrac{x^2}{9} + \dfrac{y^2}{25} = 1$

4. Sketch each hyperbola and locate its foci and asymptotes.

   a) $\dfrac{x^2}{25} - \dfrac{y^2}{9} = 1$ b) $\dfrac{y^2}{9} - \dfrac{x^2}{25} = 1$ c) $x^2 - y^2 = 1$

5. Graph the curve $xy = 1$. What type of conic is this? What will the equation of this curve become if it is rotated 45° clockwise?

6. Graph the three curves below on one coordinate sytem.

   a) $\sqrt{x} + \sqrt{y} = 1$ b) $\sqrt{x} - \sqrt{y} = 1$

   c) $\sqrt{y} - \sqrt{x} = 1$

   Can you argue that these three curves together form a parabola?

Index

ABCDE79876543210

Answers
to Selected Exercises

Chapter 1

Exercises 1–1

1. a) True   c) False   e) True   f) True   h) False   j) True   m) True

3. a) True   c) True   e) True

5. a) True   c) True   e) True   g) True   i) True   k) False   m) True
   o) True   *true*

Exercises 1–2

1. 11

3. When a variable occurs more than once in a sentence it must be replaced by the same element from the replacement set.

5. His claim is false. Different variables may be replaced by the same element from the replacement set.

7. a) 1   c) 1   e) 11   g) 66   i) 3

Exercises 1–3

1. 7, 13   3. Watch $90, ring $50

5. 14, 7   7. 6

9. This diagram would not be helpful if the sum is to be four times the difference, because then $\overline{AC}$ would have to be divided into 3 congruent segments; hence

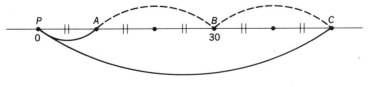

the point B (midpoint of $\overline{AC}$) would not be an endpoint of the congruent segments.

11. a) $r = 35¢$ c) $r = \$2.50$
        $s = 25¢$ $s = 50¢$
                         $u = 30¢$
                         $v = 20¢$

Exercises 1–4

1. c) 5   e) 3   g) $a = 18, b = 9$ i) $a = 9, b = 4$ k) $a = 9, b = 4$
3. a) $a = 7, b = 3, c = 2$ c) $a = 3, b = 5$ e) $a = 4, b = 2, c = 1$
   g) $a = 46, b = 5, c = 4$
5. a) $y < 7$ c) $x < y$ e) $y < 9$
7. 127

Exercises 1–5

1. a) $a^2 + ab + ab + b^2$
   b) $a^2 + ab + ac + ab + b^2 + bc + ac + bc + c^2$
   $$= a^2 + b^2 + c^2 + 2ab + 2ac + 2bc$$
3. In the 2×2 array of dots, we see along the three arrows: 1 dot, 2 dots, 1 dot. In the 4×4 array of dots, we see along the arrows: 1 dot, 2 dots, 3 dots, 4 dots, 3 dots, 2 dots, 1 dot.
5. The numbers in the box are obtained by multiplying the numbers in the cells of Exercise 4 which add up to $1^2, 2^2, 3^2$, and 4^2 by 1, 2, 3, and 4, respectively.
7. In the first case, the numbers are paired up in such a way that each sum is 11 and there are 5 pairs. In the latter case, the pairs add up to 101, and there are 50 pairs.
9. The numbers are paired in such a way that the sum of a pair is $(n + 1)$. There are $(n - 1)/2$ such pairs. Since n is odd, there is one number not paired up (the one in the "middle") $(n + 1)/2$; hence a formula is

$$\left(\frac{n-1}{2}\right)\left((n+1)\right) + \frac{n+1}{2}.$$

Exercises 1–6

1. a) 4   b) 4   c) 5   d) 5
3. a) 5   c) 9   e) 5   g) 1
5. a) $12, -2$ c) $11, -3$
7. a) $\{x \in I \mid x \le 0\}$
   c) $\{1, -2\}$
   e) $\{1, 2\}$
   g) $12, -4$

Chapter 2

Exercises 2–1

1. a) Yes   b) Yes   c) 60°, 30°, 120°, 30°   d) Yes   e) Yes   f) Yes
    g) Yes (*A, D,* and *F; A, C,* and *F; B, C,* and *E; B, D,* and *E*)

3. *New lines*

$$\overleftrightarrow{AC}, \overleftrightarrow{AD}, \overleftrightarrow{BC}, \overleftrightarrow{BD}, \overleftrightarrow{CD}, \overleftrightarrow{CE}, \overleftrightarrow{CF}, \overleftrightarrow{DE}, \overleftrightarrow{DF}.$$

New circles

A center, $\overline{AF}$ radius; *B* center, $\overline{BE}$ radius; *C* center, $\overline{CA}$ radius; *C* center, $\overline{CD}$ radius; *D* center, $\overline{DA}$ radius; *D* center, $\overline{DC}$ radius; *E* center, $\overline{EA}$ radius; *E* center, $\overline{EB}$ radius; *E* center, $\overline{EC}$ radius; *E* center, $\overline{EF}$ radius; *F* center, $\overline{FA}$ radius; *F* center, $\overline{FB}$ radius; *F* center, $\overline{FC}$ radius; *F* center, $\overline{FE}$ radius.

Exercises 2–2

1. Let *P* be the given point and $\overline{AB}$ the given segment. $\overline{PE}$ is the required segment.

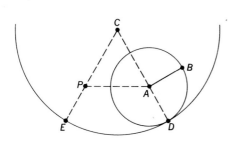

3.

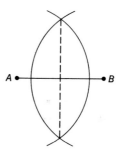

5.

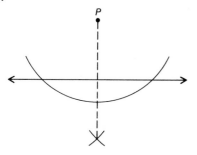

7.

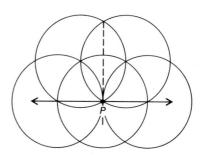

9. Given $\angle APB$

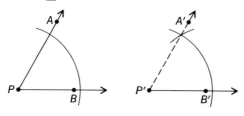

11.

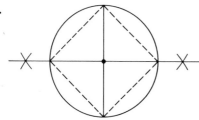

17.

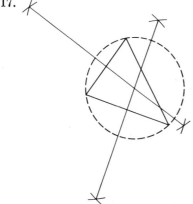

Exercises 2–4

1. 3

3. Use conjecture on triangles on same base inscribed in parallel lines. Area of
△ is $7\frac{1}{2}$.

7. $a = 3, b = 7$

Exercises 2–5

7. $7 + \sqrt{5}$ 9. 30 in. long and 4 in. wide

Exercises 2–6

3. 78 maximum surface area, 13 minimum volume.   5. 30

7. a) 8 cubic units   c) 45 cubic units   e) 30 cubic units

9. a) 1536 cubic units   c) 2490 cubic units   e) 120 cubic units

   g) 55 cubic units   *2940*

11. $9 - (\sqrt{5}/4)$

13. a) 24   c) $3^3 \cdot 24 = 648$

12.

Chapter 3

Exercises 3–1

1. a) $\overline{XY} < \overline{AB}$ b) $\overline{XY} > \overline{AB}$ c) $\overline{XY} < \overline{AB}$

3.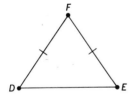

5. In $\angle DAC$ and $\angle BAC$, $\overline{AD} \cong \overline{AB}$ and $\overline{AC} \cong \overline{AC}$ but, $\overline{DC} > \overline{BC}$; therefore $\angle DAC > \angle BAC$.

7. a right angle

9. $\overline{AM} \cong \overline{MB}$

11. a) $\overline{AB} \cong \overline{A'B'}$ by (ii)   c) $\overline{AC} \cong \overline{A'C'}$ by (iii)   e) $\overline{AC} \cong \overline{DB}$ by (ii)
   f) $\overline{AB} \cong \overline{DC}$ by (iii)   i) $\angle DAE \cong \angle CAB$ by (iii)

17. In $\triangle DEF$, $\overline{DF} \cong \overline{EF}$, $\overline{DE} \cong \overline{ED}$, and $\overline{FE} \cong \overline{FD}$, therefore $\angle FDE \cong \angle FED$.

Exercises 3–2

1. $\triangle ABC \cong \triangle EDF$ 3. $\overline{ST} \cong \overline{RT}$

5. $\overline{BC} \cong \overline{AC} \Rightarrow \triangle ABC$ is isosceles.

7. $\angle ABC \cong \angle BDC$
   c) If $\triangle ABC = \triangle BCD$, then $\overline{BC} \cong \overline{CD}$, $\angle ABC \cong \angle BCD$, and $\angle BCA \cong \angle CBD$. If $\overline{BC} \cong \overline{CD}$, then $\angle CBD \cong \angle CDB$. Thus $\angle BCD \cong \angle ABC > \angle CBD \cong \angle CDB \cong \angle BCA \cong \angle BCD$, which implies that $\angle BCD > \angle BCD$, a contradiction.
   e) $\angle DBC < \angle ABC \cong \angle DCB \cong \angle BCA \cong \angle CBD \cong \angle DBC$.

13. Given $\angle ABC$.

        $\overline{BD} \cong \overline{BE}$, radii of same circle

        $\overline{DF} \cong \overline{EF}$, radii of = circle

        $\overline{BF} \cong \overline{BF}$

    $\therefore \triangle BDF \cong \triangle EFB$ by SSS.

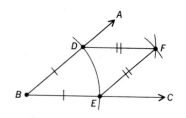

Exercises 3–3

1. a) $\angle AVF$ and $\angle FVE$ b) $\angle AVF$ and $\angle EVD$

    $\angle EVD$ and $\angle DVC$ $\angle AVF$ and $\angle AVE$

    $\angle DVB$ and $\angle BVA$

  c) $\angle AVF$ and $\angle FVD$

    $\angle EVD$ and $\angle DVB$

  d) Acute angles, $\angle FVE$ $\angle EVD$ $\angle BVC$.

  Obtuse angles, $\angle DVB$ $\angle CVE$ $\angle AVE$.

  e) $\binom{6}{2} = \dfrac{6!}{2! \, 4!} = \dfrac{6 \cdot 5}{1 \cdot 2} = 15$

  f) $\angle AVF$ and $\angle CVD$, $\angle BVC$ and $\angle FVE$, $\angle DVB$ and $\angle EVA$

3. a) $\approx \frac{1}{3}$ b) $\approx 1\frac{1}{6}$ c) $\approx \frac{2}{3}$ d) $\approx 1\frac{1}{2}$

7. Isosceles right triangles and $\cong$ b) $\cong$ Right triangles

Exercises 3–4

1. The exterior angle adjacent to the right angle is greater than the other two angles of the triangle.

3. This would violate the exterior angle theorem.

5. False. Consider a square.

9. $\overline{AD} \cong \overline{AC}$, $\angle ADC \cong \angle ACD$

    $\angle ACB > \angle ACD$

    $\angle ADB > \angle B$ (by exterior angle theorem)

  $\therefore \angle ACB > \angle ACD \cong \angle ADC > \angle B.$

Chapter 4

Exercises 4–1

1. a) $\frac{3}{3}$ c) $\frac{4}{2}$ e) $\frac{3}{2}$ g) $-\frac{3}{2}$ i) 0

3. The slope of a line which is almost vertical is very large in absolute value.

5. a) $\frac{2}{3}$ c) $-\frac{3}{2}$ e) Undefined

Exercises 4–2

3. Slope of line on first two points is $-\frac{1}{3}$.

    Slope of line on second and third points is $\frac{6}{2}$.

    The product $(-\frac{1}{3})(\frac{6}{1}) = -1$; therefore the lines are perpendicular.

5. If $\angle PAB$ and $\angle PBA$ are both right angles, then the exterior angle at B is congruent to $\angle PAB$, but this contradicts the exterior angle theorem.

9. Orthocenter of an obtuse triangle is in the exterior of the triangle.

Exercises 4–3

1. a) $\overset{\leftrightarrow}{AD}$, $\overset{\leftrightarrow}{HE}$, and $\overset{\leftrightarrow}{GF}$

    c) $\overset{\leftrightarrow}{AE} \perp \overset{\leftrightarrow}{AD}$, $\overset{\leftrightarrow}{AB} \perp \overset{\leftrightarrow}{AD}$

    e) $\angle$ B-AE-H and $\angle DAB$

    $\angle$ A-BF-G and $\angle ABC$

    $\angle$ H-DC-A and $\angle GCB$

3. a) $\cong$ c) $>$

5. No. Line m must be perpendicular to at least two lines in α to make β perpendicular to α.

7. a) 4

    c) Planes determined by pairs of parallel lines $\overset{\leftrightarrow}{EH}$ and $\overset{\leftrightarrow}{AB}$, $\overset{\leftrightarrow}{DC}$ and $\overset{\leftrightarrow}{FG}$, $\overset{\leftrightarrow}{HC}$ and $\overset{\leftrightarrow}{AF}$, $\overset{\leftrightarrow}{GB}$ and $\overset{\leftrightarrow}{ED}$, $\overset{\leftrightarrow}{HG}$ and $\overset{\leftrightarrow}{AD}$, $\overset{\leftrightarrow}{CB}$ and $\overset{\leftrightarrow}{EF}$.

    e) 7

Exercises 4–4

1. a) $\overset{\leftrightarrow}{AD} \parallel \overset{\leftrightarrow}{HE} \parallel \overset{\leftrightarrow}{GF} \parallel \overset{\leftrightarrow}{CB}$, and parallelism is transitive.

    $\overset{\leftrightarrow}{AE} \parallel \overset{\leftrightarrow}{DH} \parallel \overset{\leftrightarrow}{CG} \parallel \overset{\leftrightarrow}{BF}$, and parallelism is transitive.

    $\overset{\leftrightarrow}{AB} \parallel \overset{\leftrightarrow}{EF} \parallel \overset{\leftrightarrow}{HG} \parallel \overset{\leftrightarrow}{DC}$, and parallelism is transitive.

    c) $\overset{\leftrightarrow}{AD}$ and $\overset{\leftrightarrow}{EF}$, $\overset{\leftrightarrow}{AD}$ and $\overset{\leftrightarrow}{HG}$, $\overset{\leftrightarrow}{AD}$ and $\overset{\leftrightarrow}{CG}$, $\overset{\leftrightarrow}{AD}$ and $\overset{\leftrightarrow}{BF}$.

3. a) Yes   c) No

5. Skew lines.

7. a) $m \parallel l$ b) $m \parallel l$

11. a) $m \nparallel n$ c) Yes

13. Parallel planes

15. Right angles

17. Nonintersecting and coplanar; nonintersecting and noncoplanar; mutually intersecting and coplanar; concurrent and coplanar; concurrent and noncoplanar; one pair only intersecting and noncoplanar, etc.

Exercises 4–5

1. a) Assume Euclid's fifth postulate. Given $l \parallel m$, and t, a transversal of l and m. Assume the sum of the interior angles on one side of t are not supplementary. Then the sum of the interior angles on one side must be less than two right

angles. It follows from Euclid's fifth postulate that l and m intersect on that side of t. But this is false, since $l \parallel m$; therefore the assumption must be false.

b) Assume the theorem of Exercise (a). Let $\angle 1 + \angle 2 <$ two right angles. If l and m do not intersect, then $\angle 1 + \angle 2 =$ two right angles by the theorem of Exercise (a). This contradicts the fact that $\angle 1 + \angle 2 <$ two right angles; therefore l and m intersect. If l and m intersect on the side of t with $\angle$'s 3 and 4, then by the exterior angle theorem $\angle 2 > \angle 3$ and $\angle 1 > \angle 4$, which implies that $\angle 1 + \angle 2 >$ two right angles.

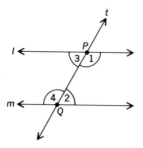

3. a) Assume Euclid's fifth postulate. Let l and m be parallel, and let n intersect l at P. Choose Q on m, and let t be the line on P and Q. Line t is a transversal of l and m and of n and m. One of the interior angles at P with respect to n and m is less than the interior at P with respect to l and m. Since the sum of both pairs of interior angles on the same side of t with respect to l and m is two right angles, the sum of one pair of interior angles on the same side of t with respect to n and m is less than two right angles. Therefore by Euclid's postulate, n and m must intersect on that side of t.

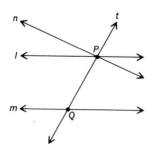

b) Assume the theorem of Exercise 3. Let l and m be cut by a transversal t such that $\angle 1 + \angle 2 < 2$ right angles. Construct $\angle 3$ at P such that $\angle 2 + \angle 3 = 2$ right angles, and the ray of l which is a side of $\angle 1$ lies in the interior of $\angle 3$. Lines n and m must be parallel. By the theorem of Exercise 3, l must intersect m. To show that the lines intersect on the required side, apply the exterior angle theorem.

Exercises 4–7

1. a) Parallelogram   c) Parallelogram   e) Trapezoid

3. Supplementary

5. a) SSS   b) Corresponding angles of $\cong$ triangles.
   c) If alternate interior angles are $\cong$ when two lines are cut by a transversal, then the lines are parallel.
   d) Opposite sides are parallel.

7. Exercise 4: If the diagonals of a convex quadrilateral form two pairs of congruent angles at opposite vertices on opposite sides of the diagonal, then the quadrilateral is a parallelogram.

9. a) SAS   b) Opposite sides of the quadrilateral are congruent.

11. a) The lines are parallel.   b) Yes.   c) Length of common perpendicular.

13. All angles are right angles.

15. The diagonals of a square are perpendicular, bisect each other, and are congruent.

Exercises 4–8

1. a) $\triangle ABC$, $\triangle A'B'C'$ b) AA', BB', CC'
   c) $ABB'A'$, $BCC'B'$, $CAA'C'$ d) Rectangles

3. Congruent

Exercises 4–9

1. Opposite sides are parallel and congruent.

3. Triangle ABC has been dissected into four congruent triangles, one of which is $\triangle ANM$.

5. Yes   7. Congruent   9. Opposite sides congruent and parallel

Exercises 4–10

3. a) 8   b) 4   d) 27

7. a) 84   c) 89.46   e) $12\sqrt{10}$ *No*

9. $288\sqrt{2}$ cubic units, $144\sqrt{3}$ square units

Exercises 4–11

5. a) $2(180°)$ b) $3(180°)$ c) $4(180°)$ d) $(n - 2)(180°)$

7. a) 7 straight angles   b) 5 straight angles   c) 2 straight angles
   d) 2 straight angles

9. a) $m \angle ECB = 45°$ $m \angle ABC = 25°$ $m \angle CDE = 95°$ b) N̶o̶ yes.

11. 1 straight angle

Chapter 5

Exercises 5–1

1. b) I^- is closed under addition, but not all elements have inverses.
   c) I^- is not closed under multiplication.

⨯ 2. a) E is closed under addition, and each element in E has an inverse in E.
   b) Yes
   c) The multiplicative inverse of 2 is not in E.

3. a) O is not closed under addition.
   b) 3 has no multiplicative inverse in 0.

4. Q^+ is closed under multiplication, and each element of Q^+ has a reciprocal in Q^+.

5. a) $1 \oplus 2 = 4, 2 \oplus 1 = 5$
   b) $(1 \oplus 1) \oplus 1 = 3 \oplus 1 = 7, 1 \oplus (1 \oplus 1) = 1 \oplus 3 = 5$
   c) If e were an identity element for the operation $\oplus$, then for every $a \in I$ we would have
   $$a \oplus e = e \oplus a \quad \text{and} \quad 2a + e = 2e + a \quad \text{and} \quad e = a.$$
   This is absurd.

6. a) Yes
   b) No. $1 \oplus (2 \oplus 3) = 22, (1 \oplus 2) \oplus 3 = 18$
   c) Yes. $a \cdot (b \oplus c) = a \cdot (2b + 2c) = 2ab + 2ac$
   $(a \cdot b) \oplus (a \cdot c) = 2ab + 2ac$
   d) No

7. a) 6   b) 6   c) 7   d) 7   e) 4   f) 4   g) 1   h) 1
   i) -3 j) 7   k) 7   l) 6   m) 6

8. a) Yes. $a + b - 1 = b + a - 1$
   b) Yes. $(a + b) + c = a + b + c - 1$
   $a + (b + c) = a + b + c - 1$
   c) Yes. The identity element for $\oplus$ is 1. $a + 1 = a + 1 - 1 = a$.
   d) The inverse of 5 relative to $\oplus$ is -3. $5 \oplus (-3) = 1$.
   e) Yes   f) Yes   g) Yes

(handwritten margin note: Not sufficient)

h) Yes, the identity element for $\odot$ is 2.

$a \odot 2 = 2a - a - 2 + 2 = a.$

9. a) No. $\quad \frac{1}{2} + \frac{-1}{2} = \frac{0}{2} = \frac{0}{1}$ $\quad$ b) Yes

c) No. The set contains neither $0/1$ nor $1/1$.

10. a) Yes $\quad$ b) Yes

c) A group under addition, but not under multiplication. The reciprocal of $\frac{2}{3}$ is not in the set.

11. We must show that $(a + b) + (\bar{a} + \bar{b}) = 0$.

$(a + b) + (\bar{a} + \bar{b}) = (a + \bar{a}) + (b + \bar{b}) = 0 + 0 = 0.$

12. We must show that $(r \cdot s) \cdot (r^{-1} \cdot s^{-1}) = 1$.

$(r \cdot s) \cdot (r^{-1} \cdot s^{-1}) = (r \cdot r^{-1}) \cdot (s \cdot s^{-1}) = 1 \cdot 1 = 1.$

13. a) No. $\quad \frac{1}{3} + \frac{1}{3} = \frac{2}{3}$

b) Yes

c) Not under addition, but it is a group under multiplication.

14. a) 3 $\quad$ b) $\frac{5}{4}$ $\quad$ c) $\frac{-3}{7}$ $\quad$ d) $\frac{-13}{5}$ $\quad$ e) $-\sqrt{5}$ $\quad$ f) $-3 + \sqrt{2}$

g) $\frac{-1}{\sqrt{7}}$ $\quad$ h) $-\left(\frac{1}{\sqrt{2}} + \frac{1}{\sqrt{3}}\right)$ $\quad$ i) $1 - i$ $\quad$ j) i $\quad$ k) $-4 + \sqrt{-7}$

l) $\frac{i - 3}{i + 5}$

(Answers are not unique expressions)

15. a) $\frac{-1}{3}$ $\quad$ b) $\frac{-4}{5}$ $\quad$ c) $\frac{7}{3}$ $\quad$ d) $\frac{5}{13}$ $\quad$ e) $\frac{1}{\sqrt{5}}$ $\quad$ f) $\frac{1}{3 - \sqrt{2}}$ $\quad$ g) $\sqrt{7}$

h) $\frac{1}{1/\sqrt{2} + 1/\sqrt{3}}$ $\quad$ i) $\frac{1}{i - 1}$ $\quad$ j) i $\quad$ k) $\frac{1}{4 - \sqrt{-7}}$ $\quad$ l) $\frac{i + 5}{3 - i}$

16. $b = c$

17. We wish to define subtraction so that if $d > a$, the symbol $d - a$ can be used for a number that adds to a to give d.

$a + (d - a) = d$

But before we can refer to *the* number that adds to a to give d, we must prove that there cannot be two such numbers. That is, we must be sure that

if $\quad a + b = d \quad$ and $\quad a + c = d,$

then $\quad b = c.$

The cancellation law for addition enables us to make this proof.

18. $a \cdot b = a \cdot c$ and

$$a \neq 0 \Rightarrow \frac{1}{a} \cdot (a \cdot b) = \frac{1}{a} \cdot (a \cdot c) \Rightarrow \left(\frac{1}{a} \cdot a\right) b = \left(\frac{1}{a} \cdot a\right) \cdot c \Rightarrow 1 \cdot b = 1 \cdot c \Rightarrow b = c$$

19. As in Exercise 17, if a is a factor of d, we wish to speak of *the* quotient $d \div a$. This is a number that multiplied by a gives d.

$$a \cdot (d \div a) = d.$$

But before we can refer to *the* quotient, d divided by a, we must prove that there cannot be two such numbers. That is, we must be sure that

if $a \cdot b = d$ and $a \cdot c = d$ and $a \neq 0$,   then $b = c$.

The cancellation law for multiplication enables us to make this proof.

20. The element a does not have an inverse relative to $*$.

Exercises 5–2

1. a) $M_r M_s(x) = M_r(sx) = (rs)x = s(rx) = M_s(rx) = M_s M_r(x)$.
   Hence

   $$M_r M_s = M_s M_r.$$

   b) $A_r A_s = A_{r+s}, \ M_r M_s = M_{r \cdot s}$
   c) $(A_r A_s)A_t = A_{(r+s)+t} = A_{r+(s+t)} = A_r(A_s A_t)$
   d) $A_r A_0 = A_r, \ M_r M_1 = M_r$; so A_0 and M_1 are the identity elements.
   e) $A_r A_{-r} = A_0, \ M_r M_{1/r} = M_1; \ A_{4/5}^{-1} = A_{-4/5}, \ M_{2/3}^{-1} = M_{3/2}$
   f) Yes

2. In each set $\mathscr{A}$ and $\mathscr{M}$, composition is an associative operation, each set has an identity element, and each element has an inverse.

3. $A_0 = M_1$

4. a) 10   b) 12   c) 7   d) 6   e) 7   f) 2

5. a) 4   b) $\frac{5}{2}$ c) 5   d) 4   e) 0   f) 15

7. a) $x = 1$ b) $x = 10$
   c) $x = 12, \ y = 14, \ z = \frac{14}{3}$
   d) $x = 8, \ y = 6, \ z = 38$
   e) $x = s/r$

8. a) Students soon observe that instead of adding 6 and then adding 4 they can simply add 10. In our notation they have seen that $A_4 A_6 = A_{10}$.
   b) The teacher calls out a number. Students are to first multiply by 2 and then multiply this answer by 5. They soon multiply by 10.
   c) Students recognize that subtracting 4 "undoes" the effect of adding 4. In our notation they recognize that $A_4^{-1} = A_{-4}$ (or S_4).

d) Students will see that to reverse the operations they must first divide by 2 and then subtract 5.

$(M_2A_5)^{-1} = A_{-5}M_{1/2}$

e) $5[4(5x + 20) - 43] + y = 100x + y + 185$
The teacher subtracts 185 from the given result and has $100x + y$. For example, if $x = 12$ and $y = 38$, $100x + y = 12\underset{\sim}{38}$. The last two digits present the mother's age, the first digits the student's age.

9. a) 4   b) 10   c) 7   d) 12   e) 7   f) $\frac{7}{2}$

10. b) $\frac{1}{2}(x - 2) = 4$ c) $2x + 6 = 20$ d) $\frac{1}{2}x - 2 = 4$
    e) $2(x - 4) + 3 = 9$ f) $2(x + 3) - 4 = 9$

11. a) $3(x + 5) = 36$
First multiply both sides by $\frac{1}{3}$, the reciprocal of 3. Then add -5, the additive inverse of 5.

13. a) $X = M_{1/3}A_{-5}$ b) $X = M_{1/3}A_{-5}$
    c) $X = M_{1/3}A_4M_{1/2}$ d) $X = M_{1/3}A_4M_{1/2}$

14. a) $A_5M_3 = X^{-1}$ c) $M_2A_{-4}M_3 = X^{-1}$

15. $r = 24, s = 120, t = 5$

16. $A_{-5}M_{3/4}A_{-2}M_{2/3}(x) = 7$; $x = M_{3/2}A_2M_{4/3}A_5(7) = 27$

17. 31 eggs

Exercises 5–3
1. The integer line has holes in it because points between 1 and 2 are not matched with integers.

   The rational number line has holes in it because if AB is any segment on a line, congruent to the diagonal of a square of length l, then one of the points A or B will not be matched with a rational number. To prove this, one shows that for no rational number r is it true that $r^2 = 2$.

2. To exhibit the sum $2 + 3$ geometrically, one might draw a circle of radius 3, center at 2.

Exercises 5–4
1. a) Fixed point 12, stretching factor 2
   b) Fixed point -12, stretching factor 2
   c) Fixed point 6, stretching factor 3
   d) Fixed point -4, stretching factor 4
   e) Fixed point 3, stretching factor 5
   f) Fixed point $-\frac{11}{5}$, stretching factor 6

2. a) Half-turn about 1   b) Half-turn about 4
   c) Half-turn about -2

3. a) $L_2 L_1(x) = -6x + 17$

   b) $L_1 L_2(x) = -3x + 4$; $L_2 L_1(x) = 28 - 3x$

   c) Think of L_1 as the function that first multiplies by 7 and then adds 3. Since $L_2(x) = 5x - 7$, $L_1(L_2(x)) = 7(5x - 7) + 3$.

4. Let $H_1(x) = a - x$, $H_2(x) = b - x$. Then

$$H_1 H_2(x) = a - (b - x) = x + (a - b),$$

$$H_2 H_1(x) = b - (a - x) = x + (b - a).$$

Since $b - a = -(a - b)$, these are translations of equal distances in opposite directions.

Exercises 5–5

2. a) $\begin{pmatrix} 3 & 0 \\ 0 & 3 \end{pmatrix} x \rightarrow \dfrac{3x}{3} = x$

   b) $\begin{pmatrix} 6 & 12 \\ 15 & -3 \end{pmatrix} = \dfrac{6x + 12}{15x - 3} = \dfrac{2x + 4}{5x - 1} = \dfrac{4x + 8}{10x - 2} = \begin{pmatrix} 4 & 8 \\ 10 & -2 \end{pmatrix}$

3. a) $\begin{pmatrix} 1 & 2 \\ 3 & -1 \end{pmatrix}$ b) $\begin{pmatrix} -1 & 3 \\ 4 & 0 \end{pmatrix}$

   c) $\begin{pmatrix} -2 & 5 \\ 0 & 3 \end{pmatrix}$ d) $\begin{pmatrix} 3 & -2 \\ 0 & 1 \end{pmatrix}$

   e) $\begin{pmatrix} -1 & 8 \\ 0 & 1 \end{pmatrix}$ f) $\begin{pmatrix} 0 & 1 \\ 1 & 0 \end{pmatrix}$

4. c) $FG = \begin{pmatrix} -1 & -1 \\ -1 & 1 \end{pmatrix}$

   $GF = \begin{pmatrix} -1 & 1 \\ 1 & 1 \end{pmatrix}$

   d) $G^2 = \begin{pmatrix} 0 & 2 \\ -2 & 0 \end{pmatrix}$

   $G^3 = \begin{pmatrix} -2 & 2 \\ -2 & -2 \end{pmatrix}$

   $G^3 F = \begin{pmatrix} 2 & 2 \\ 2 & -2 \end{pmatrix}$

   $G^3 F(x) = \dfrac{2x + 2}{2x - 2} = \dfrac{x + 1}{x - 1} = \dfrac{-x - 1}{-x + 1} = \begin{pmatrix} -1 & -1 \\ -1 & 1 \end{pmatrix} = FG$

·	I	F	G	G^2	G^3	GF	G^2F	G^3F
I	I	F	G	G^2	G^3	GF	G^2F	G^3F
F	F	I	G^3F	G^2F	GF	G^3	G^2	G
G	G	GF	G^2	G^3	I	G^2F	G^3F	F
G^2	G^2	G^2F	G^3	I	G	G^3F	F	GF
G^3	G^3	G^3F	I	G	G^2	F	GF	G^2F
GF	GF	G	F	G^3F	G^2F	I	G^3	G^2
G^2F	G^2F	G^2	GF	F	G^3F	G	I	G^3
G^3F	G^3F	G^3	G^2F	GF	F	G^2	G	I

5. a) $\dfrac{1}{2} \xrightarrow{F} -\dfrac{1}{2} \xrightarrow{G} \dfrac{1}{3} \xrightarrow{G} 2 \xrightarrow{G} -3 \xrightarrow{F} 3 \xrightarrow{G} -2 \xrightarrow{G} -\dfrac{1}{3} \xrightarrow{G} \dfrac{1}{2}$

6. $F^2 = \begin{pmatrix} -1 & -2 \\ 0 & 1 \end{pmatrix}\begin{pmatrix} -1 & -2 \\ 0 & 1 \end{pmatrix} = \begin{pmatrix} 1 & 0 \\ 0 & 1 \end{pmatrix} = I$

$G^2 = \begin{pmatrix} 0 & 4 \\ 3 & 0 \end{pmatrix}\begin{pmatrix} 0 & 4 \\ 3 & 0 \end{pmatrix} = \begin{pmatrix} 12 & 0 \\ 0 & 12 \end{pmatrix} = I$

$FG = \begin{pmatrix} -6 & -4 \\ 3 & 0 \end{pmatrix} \qquad GF = \begin{pmatrix} 0 & 4 \\ -3 & -6 \end{pmatrix}$

$(FG)^2 = \begin{pmatrix} 24 & 24 \\ -18 & -12 \end{pmatrix} = \begin{pmatrix} 4 & 4 \\ -3 & -2 \end{pmatrix}$

$(FG)^3 = \begin{pmatrix} -12 & -16 \\ 12 & 12 \end{pmatrix} = \begin{pmatrix} -3 & -4 \\ 3 & 3 \end{pmatrix}$

$(FG)^4 = \begin{pmatrix} 6 & 12 \\ -9 & -12 \end{pmatrix} = \begin{pmatrix} 2 & 4 \\ -3 & -4 \end{pmatrix}, \qquad (FG)^5 = \begin{pmatrix} 0 & -8 \\ 6 & 12 \end{pmatrix} = \begin{pmatrix} 0 & -4 \\ 3 & 6 \end{pmatrix}$

$(FG)^6 = \begin{pmatrix} -6 & -4 \\ 3 & 0 \end{pmatrix}\begin{pmatrix} 0 & -4 \\ 3 & 6 \end{pmatrix} = \begin{pmatrix} -12 & 0 \\ 0 & -12 \end{pmatrix} = \begin{pmatrix} 1 & 0 \\ 0 & 1 \end{pmatrix} = I$

Since $(FG)(GF) = F(GG)F = FIF = FF = I$, FG and GF are inverses of each other. But $FG(FG)^5 = I$; so $(FG)^5$ is also the inverse of FG, and $GF = (FG)^5$.

Now we can see that the 12 elements generated by F and G are

$I, FG, (FG)^2, (FG)^3, (FG)^4, (FG)^5$
$FI = F, FFG = G, GFG, G(FG)^2, G(FG)^3, G(FG)^4$

7. Given that $ad - bc = 0$ and $c \neq 0$. Let $a = kc$ (in other words, $k = a/c$). Then $kcd - bc = 0$ and $b = kd$. So,

$$\begin{pmatrix} a & b \\ c & d \end{pmatrix} = \begin{pmatrix} kc & kd \\ c & d \end{pmatrix} = \frac{kcx + kd}{cx + d} = k = \frac{a}{c}$$

9. For example, $a = 2, b = 1, c = -1$ yields

$$F = \begin{pmatrix} 2 & 1 \\ -3 & -1 \end{pmatrix}, \qquad F^2 = \begin{pmatrix} 1 & 1 \\ -3 & -2 \end{pmatrix}, \qquad F^3 = \begin{pmatrix} -1 & 0 \\ 0 & -1 \end{pmatrix} = I.$$

10. Set $a = 1, b = 3, c = 4$.

11. The matrix for F is $\begin{pmatrix} 1 & -(5 + 2\sqrt{5}) \\ 1 & 1 \end{pmatrix}$.

$$F^2 = \begin{pmatrix} -(4 + 2\sqrt{5}) & -(10 + 4\sqrt{5}) \\ 2 & -(4 + 2\sqrt{5}) \end{pmatrix} = \begin{pmatrix} -(2 + \sqrt{5}) & -(5 + 2\sqrt{5}) \\ 1 & -(2 + \sqrt{5}) \end{pmatrix}, \text{ etc.}$$

12. a) $\begin{pmatrix} 3 & 11 \\ 7 & -3 \end{pmatrix}^2 = I$ b) $\begin{pmatrix} 1 & 1 \\ -7 & -3 \end{pmatrix}^3 = I$

    $(3 - 3)^2 = 0 \cdot (-86)$ $(1 - 3)^2 = 1 \cdot 4$

  c) $\begin{pmatrix} 1 & 1 \\ -1 & 1 \end{pmatrix}^4 = I$ d) $\begin{pmatrix} 0 & 1 \\ -3 & 3 \end{pmatrix}^6 = I$

     $(1 + 1)^2 = 2 \cdot 2$ $(0 + 3)^2 = 3 \cdot 3$

13. For this problem we really need to know that if $AB = I$, then $BA = I$. This can be shown by a straight computation:

If $\begin{pmatrix} a & b \\ c & d \end{pmatrix}\begin{pmatrix} e & f \\ g & h \end{pmatrix} = \begin{pmatrix} 1 & 0 \\ 0 & 1 \end{pmatrix}$,   then $\begin{pmatrix} e & f \\ g & h \end{pmatrix}\begin{pmatrix} a & b \\ c & d \end{pmatrix} = \begin{pmatrix} 1 & 0 \\ 0 & 1 \end{pmatrix}$.

But the algebra is tricky. If we think of A and B as transformations on the number line, we find that the result is intuitively obvious.

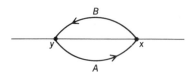

If $B(x) = y$ and $A(y) = x$ for every point x, then $AB(x) = x$ and $AB = I$. Now, clearly, $BA(y) = y$ for every point y that can be reached by B. That is, if there is a point x such that $B(x) = y$, then $BA(y) = x$. It follows that $BA = I$.

Accepting this result, we can say that if $(RS)^3 = I$, then $R(SRSRS) = I$ and $(SRSRS)R = (SR)^3 = I$.

14. a) (i) $F = \begin{pmatrix} 0 & 1 \\ 1 & 1 \end{pmatrix}$, $R = \begin{pmatrix} 0 & 1 \\ 1 & 0 \end{pmatrix}$, $A_1 = \begin{pmatrix} 1 & 1 \\ 0 & 1 \end{pmatrix}$,

$RA_1 = \begin{pmatrix} 0 & 1 \\ 1 & 0 \end{pmatrix}\begin{pmatrix} 1 & 1 \\ 0 & 1 \end{pmatrix} = \begin{pmatrix} 0 & 1 \\ 1 & 1 \end{pmatrix} = F$

(ii) $\begin{pmatrix} 1 & 2 \\ 0 & 1 \end{pmatrix}\begin{pmatrix} 3 & 0 \\ 0 & 1 \end{pmatrix}\begin{pmatrix} 0 & 1 \\ 1 & 0 \end{pmatrix}\begin{pmatrix} 1 & 2 \\ 0 & 1 \end{pmatrix} = \begin{pmatrix} 3 & 2 \\ 0 & 1 \end{pmatrix}\begin{pmatrix} 0 & 1 \\ 1 & 2 \end{pmatrix}$

$= \begin{pmatrix} 2 & 7 \\ 1 & 2 \end{pmatrix} = F$

RA_1 represents the operation of first adding 1, then taking the reciprocal. The inverse of this operation is to first take the reciprocal and then subtract 1 (add -1).

$(RA_1)^{-1} = A_{-1}R$

For $F = A_2 M_3 RA_2$, $F^{-1} = A_{-2} RM_{1/3} A_{-2}$

$F^{-1} = \begin{pmatrix} 1 & -2 \\ 0 & 1 \end{pmatrix}\begin{pmatrix} 0 & 1 \\ 1 & 0 \end{pmatrix}\begin{pmatrix} 1 & 0 \\ 0 & 3 \end{pmatrix}\begin{pmatrix} 1 & -2 \\ 0 & 1 \end{pmatrix} = \begin{pmatrix} -2 & 7 \\ 1 & -2 \end{pmatrix}$

b) $\dfrac{2x + 3}{x - 2} = 2 + \dfrac{7}{x - 2} = A_2 M_7 RA_{-2}$

The inverse is $A_2 RM_{1/7} A_{-2}$. But $RM_{1/7} = M_7 R$; so the inverse is

$A_2 M_7 RA_{-2} = \dfrac{2x + 3}{x - 2}$.

$\begin{pmatrix} 2 & 3 \\ 1 & -2 \end{pmatrix} = \begin{pmatrix} 1 & 2 \\ 0 & 1 \end{pmatrix}\begin{pmatrix} 7 & 0 \\ 0 & 1 \end{pmatrix}\begin{pmatrix} 0 & 1 \\ 1 & 0 \end{pmatrix}\begin{pmatrix} 1 & -2 \\ 0 & 1 \end{pmatrix}$

c) $\dfrac{ax + b}{cx + d} = \dfrac{a}{c} + \dfrac{b - (ad/c)}{cx + d} = A_{a/c} M_{(bc - ad)/c} RA_d M_c$

$\begin{pmatrix} a & b \\ c & d \end{pmatrix} = \begin{pmatrix} 1 & a/c \\ 0 & 1 \end{pmatrix}\begin{pmatrix} (bc - ad)/c & 0 \\ 0 & 1 \end{pmatrix}\begin{pmatrix} 0 & 1 \\ 1 & 0 \end{pmatrix}\begin{pmatrix} 1 & d/c \\ 0 & 1 \end{pmatrix}\begin{pmatrix} c & 0 \\ 0 & 1 \end{pmatrix}$

15. $\begin{pmatrix} 1 & 1 \\ 0 & 1 \end{pmatrix}^3 = \begin{pmatrix} 1 & 3 \\ 0 & 1 \end{pmatrix}$, $\begin{pmatrix} 1 & 1 \\ 0 & 1 \end{pmatrix}^n = \begin{pmatrix} 1 & n \\ 0 & 1 \end{pmatrix}$

Exercises 5–6

2. The group is commutative. $HV = VH = R$, $RH = V$, and $RV = H$.

3. The symmetries a, b, c can be visualized as reflections in the three altitudes (medians also). The symmetries R and R^2 are rotations of 120° and 240° about the center of gravity (intersection of the medians). The group is not commutative. $Ra = b$, but $aR = c$. Note that we have interpreted Ra as the entry in the *row* headed R and the *column* headed a. Since we use the function notation, $Ra(1) = R(a(1)) = R(2) = 3$, we compute right to left.

$$Ra = \begin{pmatrix} 1 & 2 & 3 \\ 2 & 3 & 1 \end{pmatrix} \begin{pmatrix} 1 & 2 & 3 \\ \underline{2} & 1 & 3 \end{pmatrix} = \begin{pmatrix} 1 & - & - \\ 3 & - & - \end{pmatrix}$$

4. $I = \begin{pmatrix} 1 & 2 & 3 & 4 \\ 1 & 2 & 3 & 4 \end{pmatrix}$

$$R = \begin{pmatrix} 1 & 2 & 3 & 4 \\ 2 & 3 & 4 & 1 \end{pmatrix}, \qquad H = \begin{pmatrix} 1 & 2 & 3 & 4 \\ 4 & 3 & 2 & 1 \end{pmatrix}$$

$$D_1 = \begin{pmatrix} 1 & 2 & 3 & 4 \\ 1 & 4 & 3 & 2 \end{pmatrix}, \qquad \ldots$$

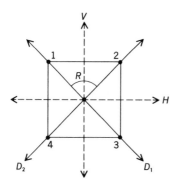

Note that $VD_1 = R$; $D_1V = R^3$; $D_1H = R$; $HD_1 = R^3$; $HD_2 = R$; $D_2H = R^3$; etc.

5. The 12 permutations consist of the identity, rotations of 120° and 240° about each of the four altitudes, and the 180° rotations around the three lines joining midpoints of opposite sides. Permutation symbols are

$$\begin{pmatrix} 1 & 2 & 3 & 4 \\ 1 & 2 & 3 & 4 \end{pmatrix}, \quad \begin{pmatrix} 1 & 2 & 3 & 4 \\ 1 & 3 & 4 & 2 \end{pmatrix}, \quad \begin{pmatrix} 1 & 2 & 3 & 4 \\ 1 & 4 & 2 & 3 \end{pmatrix},$$

$$\begin{pmatrix} 1 & 2 & 3 & 4 \\ 3 & 2 & 4 & 1 \end{pmatrix}, \quad \begin{pmatrix} 1 & 2 & 3 & 4 \\ 4 & 2 & 1 & 3 \end{pmatrix}, \quad \begin{pmatrix} 1 & 2 & 3 & 4 \\ 2 & 4 & 3 & 1 \end{pmatrix},$$

$$\begin{pmatrix} 1 & 2 & 3 & 4 \\ 4 & 1 & 3 & 2 \end{pmatrix}, \quad \begin{pmatrix} 1 & 2 & 3 & 4 \\ 2 & 3 & 1 & 4 \end{pmatrix}, \quad \begin{pmatrix} 1 & 2 & 3 & 4 \\ 3 & 1 & 2 & 4 \end{pmatrix},$$

$$\begin{pmatrix} 1 & 2 & 3 & 4 \\ 2 & 1 & 4 & 3 \end{pmatrix}, \quad \begin{pmatrix} 1 & 2 & 3 & 4 \\ 3 & 4 & 1 & 2 \end{pmatrix}, \quad \begin{pmatrix} 1 & 2 & 3 & 4 \\ 4 & 3 & 2 & 1 \end{pmatrix}.$$

6. A cube can be seen to have 24 symmetries. We can reason that since there are 8 vertices, then there are *eight* positions into which a particular vertex can be moved. After the destination of that vertex is fixed, there are *three* choices for the position of a particular adjacent vertex. As soon as the positions of two adjacent vertices are designated the symmetry is determined.

$$\begin{pmatrix} 1 & 2 & 3 & 4 & 5 & 6 & 7 & 8 \\ 8 & 4 & - & - & - & - & - & - \end{pmatrix}$$

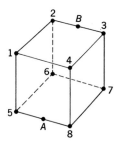

It is clear that if $1 \to 8$ and $2 \to 4$, then $5 \to 5$, $6 \to 1$, $8 \to 6$, $4 \to 7$, $7 \to 2$, and $3 \to 3$. This symmetry can be interpreted as a 120° rotation about the line through vertices 3 and 5. The 24 symmetries can be described geometrically as the identity (1), rotations of 120° and 240° around each of the four diagonals (8), rotations of 90°, 180°, and 270° around each of the three lines perpendicular to faces at their midpoints (9), and the 180° rotations about lines joining the midpoints of opposite edges (6). For example, the 180° rotation about AB as pictured is

$$\begin{pmatrix} 1 & 2 & 3 & 4 & 5 & 6 & 7 & 8 \\ 7 & 3 & 2 & 6 & 8 & 4 & 1 & 5 \end{pmatrix}.$$

If we admit as an acceptable permutation a *reflection* in a *plane* parallel to and midway between two opposite faces, a group of 48 permutations is obtained. For example, if the plane contains points A and B above, the permutation is

$$\begin{pmatrix} 1 & 2 & 3 & 4 & 5 & 6 & 7 & 8 \\ 4 & 3 & 2 & 1 & 8 & 7 & 6 & 5 \end{pmatrix}.$$

Note that this is not a *rigid motion* of the cube. A cube could not be moved in 3-dimensional space so as to fall in this position.

A regular octahedron has 24 rigid motions: 6 choices for position of a vertex and then 4 choices for an adjacent vertex. Again a reflection in a plane containing 4 vertices produces 24 more permutations.

The dodecahedron has $12 \cdot 5 = 60$ rigid motions, and the icosahedron $20 \cdot 3 = 60$ also. The relationship between the groups of motions of cube and octahedron is apparent if the figures are viewed as inscribed in a cube. Each figure is the dual of the other. A line connecting *opposite vertices* of one figure is perpendicular to *opposite faces* of the other. Dodecahedron and icosahedron are similarly related.

Chapter 6

Exercises 6–1

1. a) 3   b) 3   c) 3   d) 3   e) 0   f) 0   g) 4   h) 4

2. a) 3   b) 1   c) 4   d) 0

3. a) 5   b) 4   c) 3   d) 5
   e) No solution   f) No solution   g) 0 and 3   h) 2 and 5
   i) 1, 3, and 5   j) 2, 4   k) 3   l) 2, 4   n) 1, 2, 4, 5
   o) No solution   p) 1   q) 0, 1, 2, 3, 4, 5 (every number is a solution!)
   r) All numbers, same as (q)   s) All numbers   t) All numbers
   u) 0, 1, 3, 4   v) All numbers

4. a) 1, 3, 5   b) 2, 5   c) 2, 5   d) 0, 2, 4   e) 2, 5   f) 1, 3, 5

5. e) $(5x + 4)(5x + 2)$ f) $x^2 + x = (x + 3)(x + 4) = (5x + 3)(5x + 2)$
   k) $x^3 + 5x = x(x + 2)(x + 3) = x(x + 1)(x + 5) = (x + 3)(x + 4)(x + 5)$

6. a) 3   b) 3   c) 4   d) 4

7. a) 2   b) 3   c) 4   d) 0

8. a) 1   b) 3   c) 2   d) 4

9. a) 4   b) 4   c) 4   d) 1   e) 1   f) 4

10. a) 1, 4   b) None   c) None   d) 2, 3   e) 0   f) 3, 4   g) 4
    h) 0, 1   i) 2, 3   j) 1, 3   k) 1, 2   l) 2   m) None   n) None
    o) None

11. If $a \cdot b = 0$, then either $a = 0$ or $b = 0$, for if $a \neq 0$, a has a reciprocal x such that $x \cdot a = 1$. Then

$$a \cdot b = 0 \Rightarrow x \cdot a \cdot b = x \cdot 0 \Rightarrow b = 0.$$

12. a) 3, 2   c) 2, 1   d) 2   e) 2, 3   f) 0, 1, 3

14. $x^2 + 1 = (x + 2)(x + 3) = 4 \cdot 4(x + 2)(x + 3) = 4(x + 2) \cdot 4(x + 3)$
    $\qquad = (4x + 3)(4x + 2)$

15. a) $(x + 1)(x + 2)$, $(4x + 4)(4x + 3)$

16. a) $a = 1, b = 4, c = 3$ b) $a = 1, b = 4, c = 3$
   c) $a = 4, b = 2, c = 3$ d) $a = 2, b = 4, c = 1$

18. At step 3, if $(x + 1)(x + 3) = 0$, then $x + 1 = 0$ or $x + 3 = 0$.

19. a) $3x^2 + 4x + 1 = 0 \Rightarrow x^2 + 3x + 2 = 0 \Rightarrow (x + 1)(x + 2) = 0$, $x = 4$ or 3.

21. Substituting 4 for x, we get
$(4 + 1)(4^2 + 4 + 3) + 3 = 0 + 3 = 3.$

Exercises 6–2

1. a) $\frac{1}{2} + \frac{1}{2} = 1$, $\frac{1}{2} + \frac{1}{2} = 3 + 3 = 1$
   b) $\frac{1}{4} + \frac{1}{4} = \frac{1}{2} = 3$, $\frac{1}{4} + \frac{1}{4} = 4 + 4 = 3$
   c) $1 + \frac{1}{3} = \frac{4}{3} = 4 \div 3 = 3$, $1 + \frac{1}{3} = 1 + 2 = 3$

2. a) 3   b) 1   c) 1   d) 3   e) 3   f) 4

3. a) Since

$$\frac{b + a}{ab} = (b + a) \div ab, \qquad ab \cdot \frac{b + a}{ab} = b + a.$$

But also

$$ab \cdot \left(\frac{1}{a} + \frac{1}{b}\right) = ab \cdot \frac{1}{a} + ab \cdot \frac{1}{b} = b + a.$$

Since

$$ab\left(\frac{1}{a} + \frac{1}{b}\right) = ab \cdot \left(\frac{b + a}{ab}\right),$$

we can multiply both sides of this equation by the reciprocal of ab and get the result

$$\frac{1}{a} + \frac{1}{b} = \frac{b + a}{ab}.$$

4. a) Since $(a + b) + [(-a) + (-b)] = 0$, $-a + (-b) = -(a + b)$.
   b) $a - (b - c) = a - (b + (-c)) = a + (-(b + (-c)))$. Using part (a), we find that

$$a + (-b) + (-(-c)) = a - b + c.$$

   c) Since $ab + a(-b) = a(b + (-b)) = a \cdot 0 = 0$, $a(-b)$ is the negative of ab.
   d) Since $a(-b) + (-a)(-b) = (a + (-a)) \cdot (-b) = 0 \cdot (-b) = 0$, $(-a)(-b)$ is the negative of $-(ab)$, hence is ab.
   e) $0 - a = 0 + (-a) = -a$.

5. a) $\frac{a}{b} = \frac{c}{d} \Rightarrow d \cdot b \cdot \frac{a}{b} = b \cdot d \cdot \frac{c}{d} \Rightarrow da = bc$

b) $ad = bc \Rightarrow ad \cdot \dfrac{1}{d}\dfrac{1}{b} = \dfrac{1}{d}\dfrac{1}{b} \cdot bc \Rightarrow \dfrac{a}{b} = \dfrac{c}{d}$

6. a) Each number is the negative of a/b. For example,

$$\frac{a}{b} + \frac{-a}{b} = a \cdot \frac{1}{b} + (-a) \cdot \frac{1}{b} = (a + (-a)) \cdot \frac{1}{b} = 0.$$

b) $\dfrac{0}{a} = 0 \cdot \dfrac{1}{a} = 0$ c) $\dfrac{a}{1} = a \cdot \dfrac{1}{1} = a \cdot 1 = a.$

Exercises 6–3

1. a) $f + g = 3x + 4$ b) $f \cdot g = 2x^2 + 7x + 3$
   $[f + g](5) = 3 \cdot 5 + 4 = 19,\ f(5) + g(5) = 11 + 8 = 19$
   $f \cdot g: 5 \rightarrow 2 \cdot 5^2 + 7 \cdot 5 + 3 = 88,\ f(5) \cdot g(5) = 11 \cdot 8 = 88$

2. a) $f(4) = 11,\ g(4) = -11$
   $f + g: a \rightarrow a^2 - 2a + 3 - a^2 + 2a - 3 = 0$

3. Since $\bar{0}: x \rightarrow 0$, $f + \bar{0}: x \rightarrow f(x) + 0 = f(x)$. This function $\bar{0}$ is the only identity element for addition, for if $\bar{0}_1$ is an identity element, then

$$\bar{0}_1 = \bar{0}_1 + \bar{0} = \bar{0}.$$

4. $f + g: a \rightarrow 0 \cdot a^2 + 0 \cdot a + 1 = 1$

5. $f \cdot \bar{1}: x \rightarrow f(x) \cdot \bar{1}(x) = f(x) \cdot 1 = f(x)$.   As in Exercise 3, uniqueness may be proved.

6. b) $(f + g)(x) = f(x) + g(x) = g(x) + f(x) = (g + f)(x)$. Answers (e) and (g) are results of Exercises 3 and 5.
   f) If $ax^n + bx^{n-1} + \cdots + l$ is any polynomial, then

$$(-a)x^n + (-b)x^{n-1} + \cdots + (-l)$$

is a negative of this polynomial. No polynomial can have two different negatives, for

$$f + (-f) = \bar{0} \quad \text{and} \quad f + (f_1) = \bar{0} \Rightarrow f + (-f) = f + f_1,$$

and adding $-f$ to both sides, $-f = f_1$.

Exercises 6–4

1. The coefficient of the x^3 term in any third-degree polynomial expression is either 1 or 2. But there are three choices, 0, 1, or 2, for each of the other three coefficients. $2 \cdot 3^3 = 54$.

   Since the polynomial expressions x^3 and x denote the same polynomial function, two polynomial expressions with difference $x^3 - x$ name the same function. For example,

$$x^3 + 3x + 1 = 2x^3 + 2x + 1, \qquad 2x^3 + x^2 + 2 = x^3 + x^2 + x + 2.$$

This groups the 54 expressions into 27 pairs, with each pair naming the same function and different pairs naming different functions.

2. See answer for Exercise 1.

3. Repeatedly replacing x^3 by x, $2x + 1$.   No , X+1

4. $(x^2 + 1) \cdot (x^2 + 1) = x^4 + 2x^2 + 1 = x^2 + 2x^2 + 1 = \bar{1}$.

5. $x + 1: 2 \to 0$. Hence, if f is any polynomial, $f \cdot (x + 1): 2 \to 0$ and $f \cdot (x + 1) \neq \bar{1}$.

7. (b), (f), and (i) cannot be factored.

8. $\{2\}$

Exercises 6–5

1. $t = 2t + 1 \Rightarrow 0 = t + 1 \Rightarrow t = 2$, which is false.

2. $-(t + 2) = 2t + 1$. S is a group under addition.

4. $t(t + 1) = t^2 + t$, but $t^2 = 2t + 1$, etc.
   $2t(2t + 2) = 2 \cdot 2 \cdot t \cdot (t + 1) = 1 \cdot t \cdot (t + 1) = 1$.
   S is a field.

6. $2t + 2 = 2(t + 1)$, and $t + 1 = 2(2t + 2)$.

7. a) $2t + 2$ b) $t + 2$ c) $t + 1$ d) $t + 1$ e) $2t + 1$ f) $2t + 2$
   g) 2 h) $2t$ i) $t + 2$ j) $t + 1$ k) 1 l) 2 m) 1
   n) $2t + 1$ o) $t + 2$ p) $2t + 1$ q) $t + 1$

8. $(2t + 2)^2 + (2t + 2) + 2 = t + 2 + 2t + 2 + 2 = 0$

9. Looking at the table, we see that $(t + 2)^2 = 2$ and $(2t + 1)^2 = 2$. But $t + 1$ and $2t$ are roots of $x^2 + 2x + 2 = 0$. Using the table and noting that

$$x^2 + 2x + 2 = 0 \Leftrightarrow x^2 = x + 1,$$

we look down the diagonal for numbers x whose squares are $x + 1$.

11. $x^9 + 2x = 0$ if $x^9 = x$. We must show that the ninth power of every number in S is equal to itself. Each nonzero number in S is some power of t, and $(t^n)^8 = (t^8)^n = 1^n = 1$. So $(t^n)^9 = t^n \cdot (t^n)^8 = t^n$. Also, $0^9 = 0$, so every x in S satisfies the equation.

12. a) $2t$ b) $t + 1$ c) $t + 1$ d) 2 e) $2t$ f) $t + 1$

13. a) 1 b) 2 c) $t + 2$ d) 2

14. a) $8 \cdot 9$ b) $8 \cdot 9^2$ c) See Exercise 11.   d) 81 e) 9^9

Chapter 7

Exercises 7–1

1. Yes

3. No

5. $\angle A \cong \angle A$, $\angle AMN \cong \angle ABC$, $\angle ANM \cong \angle ACB$
$mAM = \frac{1}{2}mAB$, $mMN = \frac{1}{2}mBC$, $mAN = \frac{1}{2}mAC$

Exercises 7–2

1. (2, 4, 6, 8, 10) and (7, 14, 21, 28, 35)

3. If $a/b = c/d$, then $ad = bc$. Since $bc = cb$, $ad = cb$. If $ad = cb$, then $a/c = b/d$.

5. $a = 8$, $b = 10$, $c = 15$

7. a) $(a + b)/a = (c + d)/c = (e + f)/e$
   b) $(a - b)/b = (c - d)/d$
   c) $(a - b)/(a + b) = (c - d)/(c + d)$

9. $\begin{cases} (2 + 4 + 6 + 8 + 18) \cdot 3 = 6 + 12 + 18 + 24 + 54 \\ (3 + 6 + 9 + 12 + 27) \cdot 2 = 6 + 12 + 18 + 24 + 54 \end{cases}$

If $\dfrac{a}{a'} = \dfrac{b}{b'} = \dfrac{c}{c'} = \dfrac{d}{d'} = \dfrac{e}{e'}$,

then $\dfrac{a + b + c + d + e}{a' + b' + c' + d' + e'} = \dfrac{a}{a'}$.

11. a and b, a and c, b and c

13. a) $5, 4$ b) $7\frac{1}{2}, 6$ c) $25, 15$ d) $6\frac{1}{2}, 6$ e) $12\frac{1}{2}, 12$ f) $13\frac{1}{2}, 6\sqrt{5}$

15. a) $a = 1$ c) $e = 7\frac{1}{2}$ e) $e = 5\frac{2}{5}$ g) $b = \frac{2}{3}$ i) $a = \frac{100}{51}$

Exercises 7–3

5. a) Triangles with same base inscribed in parallel lines
   b) Equals added to equals
   c) $\overline{AE}$ and $\overline{AC}$ are the bases, and B the opposite vertex
   d) Same as (c)
   e) Triangles BAE and CDA have equal areas, and $\triangle BAC$ and $\triangle CBA$ are identical.

$$\dfrac{\triangle CDA}{\triangle CBA} = \dfrac{\triangle BAE}{\triangle BAC}$$

8. By AAA

9. $\triangle ABS \sim \triangle ACR$; $\triangle RSQ \sim \triangle RAP$.

Exercises 7–4

1. a) Yes, condition 3   b) Yes, condition 3   c) Yes, condition 3

3. a) 28   b) $\frac{100}{3}$

5. a) Yes   b) No   c) Yes   d) Yes   e) No   f) Yes   No

7. Use other pair of diagonals

9.

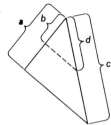

11. $\triangle ANM \sim \triangle ABC$, $m(\overline{AN}/\overline{AB}) = m(\overline{AM}/\overline{AC}) = \frac{1}{2}$, and

area $\triangle ANM = \frac{1}{4}$ area $\triangle ABC$,

since $\triangle ANM$ is congruent to the other three smaller triangles.

13. 1 to 16

15. 9 to 16

Exercises 7–5
3. 667 mi

Chapter 8

Exercises 8–1

1. a) 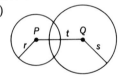   c)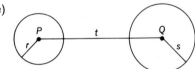

e)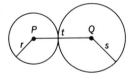

3. a) False   b) True   c) True   d) False
   e) False   f) True   g) True

5. a) Radii of the same circle are ≅.
   b) $\angle PXQ \cong \angle PYQ$ (base $\angle$s of an isosceles triangle). Hence $\triangle PQX \cong \triangle PQY$ because of hypotenuse-angle congruence theorem for right triangles.
   c) Corresponding parts of ≅ triangles.
   d) A radius $\perp$ to a chord of a circle bisects the chord.

7. The $\perp$-bisectors of the three chords are concurrent at the center of the circle. Therefore to construct the circle cirumscribing a given triangle, construct the

⊥-bisectors of the sides; the point of intersection of the ⊥-bisectors is the center of the required circle.

9. a) $\overline{AD} \cong \overline{CB}$ and $\overline{AC} \cong \overline{BD}$.

b) The end points of two distinct diameters of a circle determine a parallelogram (actually a rectangle which will be clear when inscribed angles are discussed).

c) $\triangle CPB \cong \triangle APD$ by SAS $\triangle CPA \cong \triangle BPD$ by SAS

Hence $\overline{AD} \cong \overline{CB}$ and $\overline{AC} \cong \overline{BD}$ by corresponding parts. $ACBD$ is a parallelogram because the opposite sides of the quadrilateral are congruent.

11. In a circle, the midpoints of all the chords congruent to a given chord is *a circle*.

Proof: Congruent chords are equidistant from the center of a circle. The distance is the ⊥ distance from the center to the chord, and a radius ⊥ to a chord bisects the chord. Therefore, the midpoints of congruent chords are equidistant from the center of the circle.

Exercises 8–2

1. Cans for food, drinking glasses, lamp shades

3. a) Cone   b) Circular cylinder   c) Circular cylinder

5.

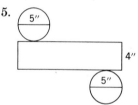

7. A closed curve congruent to the base of the cylinder

9. Right cylinder

Exercises 8–3

1. a) Pencil and a ball   b) Wirehoop and a piece of cardboard
   c) Ball and a piece of cardboard

3. a) $\overleftrightarrow{OP} \perp l$, converse of theorem stated above

   b) $\overleftrightarrow{O'P} \perp l$, same reason,

   c) O, O', P must be collinear, otherwise there would be two perpendiculars to l at P.

5. Two

7. Place the center square such that it touches the circle at C and D. The lines $\overleftrightarrow{CQ}$ and $\overleftrightarrow{DQ}$ are tangent to the circle. Since PQ bisects $\angle CQD$, it contains the center of the circle.

9. The chords are all congruent. The point of tangency is the midpoint of each chord.

11. a) Infinitely many

 b) The centers of all the circles containing a pair of distinct P, Q points all lie on the perpendicular bisector of $\overline{PQ}$, and each point of the perpendicular bisector of $\overline{PQ}$ is the center of a circle containing P and Q.

13. $\overleftrightarrow{OP} \perp \overleftrightarrow{PA}$, $\overleftrightarrow{PB}$, and $\overleftrightarrow{PC}$

 For example: O, P, and A determine a unique plane which intersects the sphere in a great circle. Since the sphere intersects π in only one point P, $\overleftrightarrow{PA}$ is tangent to the great circle at P. A radius is perpendicular to a tangent at its extremity.

Exercises 8–4

1. a) $\widehat{ACB}$ and $\widehat{ADB}$ b) $\widehat{AC}, \widehat{CB}, \widehat{BD}, \widehat{DA}$ c) $\widehat{CBA}, \widehat{BDC}, \widehat{DAB}, \widehat{ACD}$

3. $m\widehat{DC} = 100°$, $m\widehat{AB} = 100°$ $m\widehat{AD} = m\widehat{BC} = 100°$

5. a) With the vertex of the right $\angle$ as center and any radius, construct a circle intersecting the sides of the right $\angle$.

 b) With one point of intersection as center, and the same radius as in (a), draw an arc intersecting the circle in the interior of the angle at P.

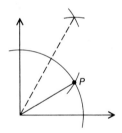

 c) Draw the segment determined by the vertex and P. The line between P and the vertex of the right $\angle$ determines a 60° angle and a 30° angle.

 d) Bisect the 60° angle.

7. The points X, Y, Z are collinear.

9. $m\angle BAD = \frac{1}{2}m\widehat{BD}$, $m\angle CAD = \frac{1}{2}m\widehat{DC}$

 $\therefore m\angle BAC = \frac{1}{2}m\widehat{BDC}$

11. The measure of an inscribed angle is $\frac{1}{2}$ the measure of its subtended arc.

13. The subtended arcs of the opposite angles of a quadrilateral inscribed in a circle comprise the complete circle. Hence, the sum of the measures of the

subtended arcs is 360°. It follows from the theorem of Exercise 11 that the sum of the measures of the opposite angles is 180°.
The opposite angles must be supplementary.

15. a) Bisect $\overline{OP}$ (M the midpoint).

b) Construct a circle with M as center and $\overline{MP}$ and $\overline{MO}$ as radii.

c) Let A, B be points of intersection with the original circle and the circle of part (b). Then $\overleftrightarrow{PA}$ and $\overleftrightarrow{PB}$ are the required tangents. To verify, note that $\angle OAP$ and $\angle OBP$ are right angles because they are inscribed in semicircles.

Exercises 8–5

1. a) $m\angle BAC = 40°$

b) Line m will not be perpendicular to $\overline{AC}$ and hence is not tangent. A line intersecting a circle in one point and not tangent must intersect the circle in a second point.

c) Alternate interior angles determined by a transversal and parallel lines

d) Inscribed angle measure theorem

e) The theorem is verified by the discussion preceding the exercises.

f) $m\angle BAC = \frac{1}{2}m\widehat{AC}$

g) The measure of an angle formed by a tangent and a chord is $\frac{1}{2}$ the measure of the arc in the interior of the angle.

3. $m\angle GAC = \frac{1}{2}[m\widehat{GBC} - m\widehat{GFE}]$
   $m\angle GAB = \frac{1}{2}[m\widehat{GB} - m\widehat{GF}]$ } By results of Exercise 2

   $m\angle BAC = m\angle GAC - m\angle GAB$

   $\qquad = \frac{1}{2}[m\widehat{BC} - m\widehat{FE}]$

The measure of an angle formed by two secants, exterior to the circle, is half the difference of the intercepted arcs.

5. $m\angle 1 = 50°$ $m\angle 2 = 50°$ $m\angle 3 = 50°$ $m\angle 4 = 25°$

Exercises 8–6

1. None of these   No

3. a) Rhombus   b) Rectangle   c) Possible

5. Adjust the compass until one can cut off 5 approximately congruent arcs on the circle. Of course this construction is not exact.

7. a) No

b) (i) If the opposite angles of a quadrilateral are supplementary, then the quadrilateral can be circumscribed. (ii) If opposite angles are not supplementary, the quadrilateral cannot be circumscribed by a circle.

9. Equilateral triangle

11. The circle of A, C, and D must have $\overline{AC}$ as a diameter since $\angle ADC$ is a right angle; therefore $\overline{AC}$ must subtend an arc of 180°. Likewise the circle on A, C, and F has AC as a diameter. The midpoint of $\overline{AC}$ is the center of both circles and the radii are equal; this implies that the circles are identical. Thus there is a circle on A, C, D, and F with $\overline{AC}$ as a diameter.

    Other cyclic quadrilaterals are $ABDE$ and $BCEF$.

13. a) Solve the following equation for n:

    $n \cdot 128\frac{4}{7} = (n-2)180,$ n represents the number of sides

        $n = 7$

    b) 9   c) 10   d) 18

15. a) A seven-pointed star

    b) No

17. 3: regular tetrahedron, regular octahedron, regular icosahedron

19. No. Three angles must meet at each vertex. Their sum must be less than 360°, so each must be less than 120°. Regular polygons of 6 sides or more have angles greater than or equal to 120°.

Exercises 8–7

1. a) 25π in.

    b) About 800

3. 8π inches; $8\pi\sqrt{2}$ inches

   5. 56π

7. a) Half the area of the circle. Since the small arcs are semicircles and the chords of the semicircles are congruent, a segment connecting the end points of the semicircles on the large circle is a diameter. The shaded and unshaded regions on opposite sides of the diameter are congruent.

    b) The line bisecting both the shaded and unshaded regions makes a 45° angle with the dotted line.

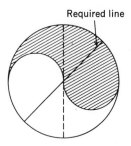

Required line

9. a) The number of diagonals of a pentagon is 3 more than for a quadrilateral; of a hexagon 4 more than for a pentagon; another side gives 5 more diagonals, then 6 more, etc.

Polygon	Number of Diagonals
4-gon	$2 = 1 + 1$
5-gon	$5 = 1 + 2 + 2$
6-gon	$9 = 1 + 2 + 3 + 3$
7-gon	$14 = 1 + 2 + 3 + 4 + 4$
8-gon	$20 = 1 + 2 + 3 + 4 + 5 + 5$
9-gon	$27 = 1 + 2 + 3 + 4 + 5 + 6 + 6$

b) 35, 44, 189   c) $\dfrac{n(n-3)}{2}$

Exercises 8–8

3. The volumes are equal.

5. Doubling the diameter increases the volume by a factor of four because the radius is squared.

Exercises 8–9

1. a) The shaded region on the sphere
   b) Radius of the sphere
   c) $\frac{1}{3}$(area of base)(radius of sphere) $= \frac{1}{3}br$
   d) $4\pi r^2$
   e) $\frac{1}{3}(4\pi r^2)r = \frac{4}{3}\pi r^3$
      Total surface area of sphere = area of all bases of "pyramids."

3. The larger sphere has a volume eight times greater than that of the smaller sphere.

Chapter 9

Exercises 9–1

1. a)

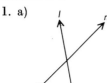

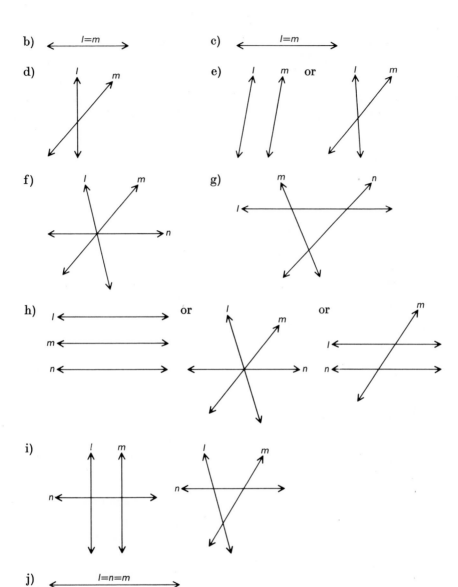

Exercises 9–2

1.

Four betweenness relations for four points: (ABC), (ABD), (ACD), (BCD).
Ten betweenness relations for five points.

3. A, B, C are not collinear.

5. $(XABC)$, $(AXBC)$, $(ABXC)$, $(ABCX)$

7. a) True   b) True   c) False   d) True   e) True

9. Let S be the set of points of the circle and its interior (circular disk). If T is the diameter $\overline{AB}$; A and A' as indicated in the diagram, then T separates S

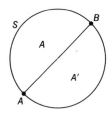

Exercises 9–3

1. a) (XPY); $P \in \overline{XY}$; $\overrightarrow{PX}$ and $\overrightarrow{PY}$ are opposite rays.

   b) $\overline{CD}$ contains a point of $\overleftrightarrow{EF}$.

   c) $\overline{RS}$ contains a point of π.

   d) $\overline{UT}$ does not contain a point of $\overleftrightarrow{MN}$.

3. $S \in l$ or $S \notin \alpha$.

4. a) $\overline{XB}$ c) $\overrightarrow{AB}$ e) $\overrightarrow{AB}$ g) $\overleftrightarrow{AB}$

5. a) $\overleftrightarrow{AB}$ c) $\emptyset$ e) $\{B\}$ g) $\overrightarrow{AB}$

7. Prove $\overline{AB} = \overline{BA}$

   i) Show $\overline{AB} \subset \overline{BA}$.

   Let $X \in \overline{AB}$; then $X = A$ or $X = B$ or $A \neq X \neq B$ and (AXB).
   This is the same as $X = B$ or $X = A$ or $B \neq X \neq A$ and (BXA). Therefore $X \in \overline{BA}$. Thus for all $X \in \overline{AB}$, X is also an element of $\overline{BA}$ which implies that $\overline{AB} \subset \overline{BA}$.

   ii) Show $\overline{BA} \subset \overline{AB}$. The proof is similar to the above argument. $\overline{AB} \subset \overline{BA}$ and $\overline{BA} \subset \overline{AB} \Rightarrow \overline{AB} = \overline{BA}$.

9. Consider rays $\overrightarrow{AB}$ and $\overrightarrow{CD}$.

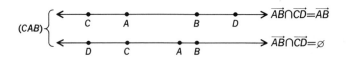

9. (*continued*)

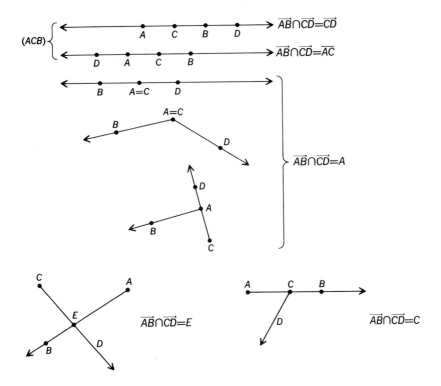

$\overrightarrow{AB} \cap \overrightarrow{CD} = \overrightarrow{CD}$

(ACB)

$\overrightarrow{AB} \cap \overrightarrow{CD} = \overline{AC}$

$\overrightarrow{AB} \cap \overrightarrow{CD} = A$

$\overrightarrow{AB} \cap \overrightarrow{CD} = E$

$\overrightarrow{AB} \cap \overrightarrow{CD} = C$

11. a) *All lines are coplanar.*

Take half-plane determined by n containing A.
Take half-plane determined by l containing B.
Take half-plane determined by m containing A.

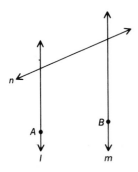

b) Take half-plane determined by l containing C.
Take half-plane determined by m containing B.
Take half-plane determined by n containing A.

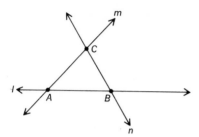

Exercises 9–4

1. A dihedral angle is the union of two half-planes and their common edge. The common edge corresponds to vertex. Acute, right, obtuse, flat. The interior of a dihedral angle, formed by half-planes α containing C and β containing D with common edge $\overline{AB}$, is the intersection of the half-space determined by plane α containing D and the half-space determined by plane of β containing C.

5. a) The union of the three angles is $\overrightarrow{AB} \cup \overrightarrow{AC} \cup \overrightarrow{BA} \cup \overrightarrow{BC} \cup \overrightarrow{CA} \cup \overrightarrow{CB}$.
   b) The intersection of the three angles is the set $\{A, B, C\}$.
   c) The intersection of the interiors of the angles is the interior of the triangle.

7.

Consider the half-line of l determined by A and containing C. If X, Y are any two points of this half-line, then $\overline{XY}$ is a subset of this half-line.

9. a)   b)   c)

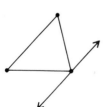 

11. The line l contains a point of $\overline{BC}$ or $\overline{AC}$.

Exercises 9–5

1. a) Circle: $0°$ i.e., parallel to α
      Parabola: $45°$
      Hyperbolas: $45° < x \le 90°$
      Ellipses: $0° < x < 45°$

   b) Single point, as shown in the figure
      Single line, $45°$ angle with α and containing V
      Two lines, $90°$ angle with α and containing V

6. If P is the point of intersection of the crease and the radius to the point of the circle which F is folded upon, then $\overline{FP} + \overline{PO}$ = radius of the circle. Thus the creases are tangents to the point ellipse.

8. Let F be folded upon X and let P be the intersection of $\overset{\leftrightarrow}{OX}$ and the crease. Then $\overline{XP} \cong \overline{FP}$ and $\overline{FP} - \overline{PO} = \overline{OX}$, radius of the circle. Hence the collection of points P is a hyperbola with P and O as foci, and the creases are tangent to the hyperbola forming the corresponding line conic.

9. Ellipse containing Q and Q'

10. Circle with fixed point as center

11. Ellipse containing Q and Q'

12. Hyperbola containing Q and Q'

13. Line ellipse

15. If l does not meet $\overline{VW}$, then the locus is an ellipse containing V and W.

    The ellipse approaches a circle when the distance from l to W (or V) is increased.

    If l meets $\overline{VW}$, then the locus is a hyperbola.

16. Points are collinear.

17. Points are collinear.

18. Lines are concurrent.

Chapter 10

Exercises 10–1

3. We did not win the baseball game *and* we did not win the track meet.

4. We did not win the ball game *or* we did not win the track meet.

5. Of course, common sense insists that statements "*p* or *q*" and "*q* or *p*" have the same meaning. Formal use of the truth tables shows equivalence, for when either one is true, so is the other.

6. a) *p* is false and *q* is false.
   b) *p* is false or *q* is false.

c) p is true and q is false.

d) q is true and p is false.

e) One is true and the other false.

7. a) True   b) False   c) True   d) True

8. Let p be the statement, "You will not wash the dishes tonight," q the statement, "You will not go to the party."

9. a) If you do not sit quietly, then I will spank you.

b) If $x \neq 2$, then $x = 4$.   c) If $x \neq 4$, then $x = 2$.

d) If I do not spank you, then you will have sat quietly.

e) If not p, then q.   f) If not p, then p.   g) If p then q.

h) If p, then not q.

10. a) He will telephone soon or I will be angry.

b) It will not turn cold or it will rain.   c) $x = 4$ or $x = 5$

d) $x \neq 4$ or $x = 5$ e) $x \neq 5$ or $x \neq 4$ f) $l \parallel m$ or $l \nparallel n$

h) (not p) or q i) (not q) or p j) p or q k) (not q) or not p

l) p or not q m) q or not p n) q or not q o) (not q) or (not q)

Exercises 10–2

1. a) The statement, (not p) or q, is true, so one of the two, (not p), q is true. But (not q) is true, so q is false. Hence, (not p) is true and p is false.

b) (not p) or (not q) is true; q is true; (not q) is false; so (not p) is true and p is false.

h) $p \Rightarrow q$ means (not p) or q. (not q) $\Rightarrow$ (not p) means (not (not q)) or (not p), that is, q or (not p).

i) (not (p or q)) means (not p) and (not q). Hence, p is false; q is false; $p \Rightarrow q$; $q \Rightarrow p$.

2. a) $p \Rightarrow q$ means (not p) or q. We are given that q is true, so p may be either true or false. We haven't enough information to decide which.

b) $p \Rightarrow q$ is true when p is false and q is true. But $q \Rightarrow p$ is not true under those conditions.

c) p could be true, q false, and r true. Then $q \Rightarrow r$ is true, and $p \Rightarrow (q \Rightarrow r)$ is true.

3. Here is a truth table proof for 1(c). We filled the table out "mechanically," using the rules.

p	q	(not q)	(not p)	(not $q \Rightarrow$ not p)
T	T	F	F	T
F	T	F	T	T
T	F	T	F	F
F	F	T	T	T

We see that *only* the fourth line of the table meets the conditions of the hypothesis. On this line, (not p) is true.

4. a) not p b) q c) not p d) not q; not p

5. $p \Rightarrow$ not q is the statement: *If* there is a line l and a point P not on l such that there are two distinct perpendicular lines m and n from P to l, *then* there is a triangle such that the sum of its angles is *not* a straight angle.

The figure suggests the proof of this theorem, $p \Rightarrow$ (not q).

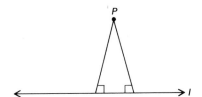

Since we have assumed that q is true, (not q) is false; and so p is false, and there is no point from which two different perpendiculars can be drawn to a line.

Exercises 10–3

1. a) Roses are not red *or* violets are not blue.
   b) Roses are not red *and* violets are not blue.
   c) Roses are red *and* violets are not blue.

2. a) There is a cat that is not black.
   b) There is a Chinese who is not short.
   c) There is a dishonest Frenchman.
   d) $\exists x \in U$, x is not a squirrel.
   e) $\exists x \in U$, x is not good or x is not brave.
   f) $\exists x \in U$, $x > 20$ and $x \geq 30$.
   g) $\forall x \in U$, if x is Chinese, then x is not tall.
   h) Every cat in the pound has a tail.
   i) $\forall x \in U$, x does not stammer.
   j) $\forall x \in U$, x is odd or $x \leq 40$.
   k) $\forall x \in U$, $\exists y \in U$, $y \neq x$ and $y \geq x$.

3. a) True   b) False   c) True   d) False   e) True   f) True
   g) False

4. a) The negation of "All roses are red and all violets are blue" is "There exists a rose that is not red or there is a violet that is not blue."

5. a) True   b) True   c) True   d) True   e) False
   f) True (if we accept the definition that a line is parallel to itself)
   g) True   h) False   i) False   j) True   k) True
   l) False (for example, ──────•────•────•──────). With the added
                                    A B C

   $\angle ABC \not\cong \angle ACB$

   condition that A, B, C are not collinear, (l) is true.

Exercises 10–4

1. The diagonals of a rhombus are perpendicular to each other, as are those of a kite-shaped figure.

   b) An isosceles triangle may not be equilateral.

   c) $3 + 5$ is even.

   d) $x = 1$ and $y = 6 \Rightarrow x + 2y = 13$

   e) $(-3)^2 > 0$ and $-3 \not> 0$

2. a) If the diagonals of a quadrilateral are not perpendicular, then the figure is not a square.

   c) If $a + b$ is odd, then it is not the case that both a and b are even.

3. Implications (a) and (d) are true.

4. a) $\forall a, b \in W, a + 2 > b + 2 \Rightarrow a > b$

   b) $m^2 = n^2 \Rightarrow m^2 - n^2 = 0 \Rightarrow (m + n)(m - n) = 0 \Rightarrow m + n = 0$ or $m - n = 0 \Rightarrow m = -n$ or $m = n$

   c) For all triangles ABC, if $\angle C > \angle B$, then $\overline{AB} > \overline{AC}$. This is the "greater angle subtends the greater side" theorem.

   d) The converse is the SAS theorem.

5. a) $\forall a, b \in W, a \neq 0,$ and $b \neq 0 \Rightarrow ab \neq 0.$

   b) For all quadrilaterals, if a quadrilateral is a rhombus then its diagonals are perpendicular.

   c) If two lines (coplanar) are not parallel, then no transversal forms a pair of congruent corresponding angles.

   d) A triangle with no pair of congruent angles is not isosceles.

   e) If circumcenter and incenter of a triangle are distinct, then the triangle is not equilateral.

   f) If a circle circumscribes quadrilateral $ABCD$, then $\angle A$ and $\angle C$ are supplementary.

Exercises 10–5

1. p only if q 2. q is necessary for p

3. p is sufficient for q

4. a) Diagonals bisect each other.

   b) The quadrilateral is a parallelogram.

5. a) The quadrilateral is a parallelogram.

   b) The diagonals bisect each other.

6. a) True   b) False

   c) False   d) True

   e) True   f) False

   g) True   h) False

   i) True   j) False

   k) False   l) True

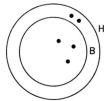

Chapter 11

Problem 1. Yes; no; no.

Problem 2. Yes; no.

Problem 3. No; no.

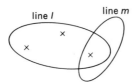

Theorem 1. Let A and B be any two points. Name the line that contains these points l. There is a point not on l. Call this point C. There is a line m containing A and C, and a line n containing B and C. Also $l \neq m$; $m \neq n$; $l \neq n$.

Theorem 2. If two lines were parallel there would be 6 points, A, B, C on one line, D, E, F on the other. Lines $\overleftrightarrow{AD}$ and $\overleftrightarrow{AE}$ would contain third points G, H, respectively. It is easy to see that $G \neq H$, and the plane contains at least 8 points, contradicting Axiom 1.

Theorem 4. If a line l had more than 3 points on it, then there would be at least 4 points on l. Name these points A, B, C, D. There is a point E not on l. Lines $\overleftrightarrow{EA}, \overleftrightarrow{EB}, \overleftrightarrow{EC}$ now contain new points bringing the total to more than 7, which contradicts Axiom 1.

Theorem 5. Four lines on a point A would yield 9 points in the plane.

Theorem 6. Let A be any point. There is a point $B \neq A$. Line $\overleftrightarrow{AB}$ is on A. There is a point $C \notin \overleftrightarrow{AB}$. Line $\overleftrightarrow{AC}$ is on A. There is a point D on line $\overleftrightarrow{BC}$ with $D \neq B$ and $D \neq C$. Line $\overleftrightarrow{AD}$ is on A, and it is easy to see that $\overleftrightarrow{AD} \neq \overleftrightarrow{AB}$ and $\overleftrightarrow{AD} \neq \overleftrightarrow{AC}$.

Theorem 7. This combines Theorems 4, 5, and 6.

Theorem 8. Let A, B, C be any 3 noncollinear points. There are 3 points D, E, F on $\overleftrightarrow{BC}, \overleftrightarrow{AC}, \overleftrightarrow{AB}$, respectively. On $\overleftrightarrow{AD}$ there is a third point G which is not equal to any of the other 6. Hence there are at least 7 points. But by Axiom 1, there are no more than 7 points, so there are exactly 7.

Theorem 9. The points of Theorem 8 yield at least 7 lines: $\{A, B, F\}, \{A, C, E\}, \{B, C, D\}, \{A, D, —\}, \{B, E, —\}, \{C, F, —\}, \{D, E, —\}$. By elimination we can see that the third point in $\{A, D, —\}$ is G, as it is in $\{B, E, —\}$ and $\{C, F, —\}$. It follows that F is in line $\{D, E, —\}$.

Note that each of the 7 lines includes 3 pairs of points. For example, $\{A, B\}, \{A, F\}, \{B, F\}$ are subsets of $\{A, B, F\}$. These 21 pairs are distinct and comprise all the $(7 \cdot 6)/2$ pairs that can be formed from 7 points. An eighth line, then, cannot include any of these pairs and must introduce at least 2 new points, contradicting Axiom 1.

Hence there are 7 and no more than 7 lines.

Problem 4. 28 triangles. There are $(7 \cdot 6 \cdot 5)/(3 \cdot 2 \cdot 1) = 35$ subsets of 3 points, and 7 of these are collinear sets.

Problem 6. No. In the figure shown, $\{D, E, F\}$ is the "strange" line.

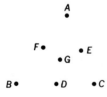

Exercises 11–2

1. Euclid: Let it be postulated that one can draw a line from any point to any point.
2. Euclid's first postulate does not assure us of this. In his definitions Euclid assumes that lines have points.
3. Equals added to equals . . . .
4. Hilbert's Postulates 4, 5, 6, and 7 state properties of betweenness that Euclid overlooked.
5. Equals subtracted from equals . . . is proved by Hilbert; Euclid's postulate that circles can be "drawn" is replaced by a definition; Euclid's postulate that all right angles are "equal" becomes a theorem for Hilbert; Euclid's postulate that things that coincide are equal is replaced by postulates characterizing congruence.

6.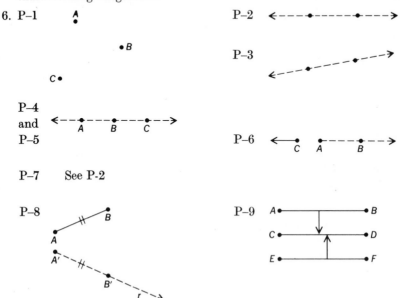

P–10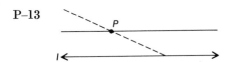

P–11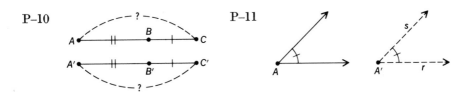

P–12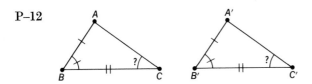

P–13

Chapter 12

Exercises 12–1

3. The function denoted by the sign "$-$" maps each whole number in W upon an element of W^-. Since this is a one-to-one, onto correspondence from W to W^-, the *inverse* of this function maps W^- one-to-one onto W. Combining the symbol $-$ with its inverse gives a function that maps $W \cup W^-$ one-to-one onto itself. $W \cup W^-$ is called the set of integers and is denoted by I. The new function is referred to by the language, "the opposite of," "the negative of," and "the additive inverse of." For example, the negative of 3 is -3, and the negative of -3 is 3.

4. a) If $x \neq y$, $2x + 1 \neq 2y + 1$, so f is 1-1.
   For each odd whole number k there is a whole number n such that $f(n) = 2n + 1 = k$, so f is onto.
   b) If $x \neq y$, $2x + 7 \neq 2y + 7$, so f is 1-1.
   But there is no $x \in W$ so that $2x + 7 = 3$, so f is not onto.
   c) If k is any odd whole number, then $k - 1$ is an even whole number, and so by the definition, $f(k - 1) = (k - 1) + 1 = k$. Hence f is onto. But $f(3) = 7$ and $f(6) = 7$, so f is not 1-1.

5. For no $x \in I$ is $f(x) = 11$. So f is not onto. Moreover, $f(-3) = f(3)$, so f is not 1-1.

7. a) Use equal distances to pair points

b)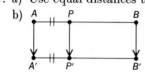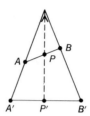

c) Match the points of each side of one triangle with those of some side of the other.

Exercises 12–2

1. a) Lines parallel to l b) No

2. We need the theorem, "If two sides of a convex quadrilateral are congruent and parallel, the figure is a parallelogram." $T(m)$ is parallel to m.

3. $T(\overline{XY}) \cong \overline{XY}$

4. $\triangle ABC \cong \triangle A'B'C'$ (by SSS)

5. a) All lines parallel to the direction of translation
   b) All planes that include a line of (a).

6. The negative of the vector that determines T

Exercises 12–3

1. No; no; no triangles; all circles with center at C; all squares with center at C.

2. $l \perp l'$

3. A 90° rotation in the opposite sense

4. a) 180° rotation about C
   b) 270° rotation
   c) The identity transformation

5. Yes

7. a) A square
   b) A square

8. A regular hexagon; A regular 36-sided figure. For repeated 17° rotations a point would have 360 images (including itself), and the points would be vertices of a regular 360-gon.

9. A rotation of a plane can be thought of as a rotation about a line perpendicular to the plane. We can visualize a rotation of 3-dimensional space about a line, and then the "movement" of points in one plane is just part of the rotation in space.

Exercises 12–4

3. If we have designated one side of the point of reflection as the *positive* side, and one as the *negative* side, these sides interchange. If we view the line from a fixed side of the line in the plane, left and right interchange.

6. a) $M_l(m) \parallel l$ b) $M_l(n) = n$
   c) $M_l(p)$ intersects l at A.

7. Line l is the perpendicular bisector of $\overline{AB}$.

8. Line l is the bisector of $\angle ABC$.

Exercises 12–5

1. a) Translation in direction of $\overrightarrow{AA'}$, with distance, $m\overline{AA'}$
   b) Rotation of 180° about the midpoint of $\overline{AA'}$
   c) Rotation of 90° about the point of intersection of $\overleftrightarrow{AB}$ and $\overleftrightarrow{A'B'}$
   d) Reflection in the line which bisects $\angle ATA'$ where T is the point of intersection of $\overleftrightarrow{AB}$ and $\overleftrightarrow{A'B'}$

3. B, C, D, E, M, N, S, T, U, V, W, Y, Z have groups of two transformations. G, J, K, L, P, Q, R have only one, the identity. Of course different writing styles affect these conclusions.

4. In order for these letters to be transformed into each other a simple writing style must be adopted. Then a 180° rotation will change p into d and a reflection will interchange b and d.

Exercises 12–6

9. Lines a and b are perpendicular.

10. a) $M_a M_b$ is a rotation unless the lines are parallel. But in this case, $M_a M_b$ would be a translation and $(M_a M_b)^5 \neq I$.
    b) Either 36° or 72° since repeating each five times yields the identity.

11. a) 23   b) 15   c) 16   d) 22   e) 61   f) 139   g) 65   h) 61

12. a) Yes. $M_a M_c$ repeated maps square 1 on "half" the squares in the column containing 1, and $M_c M_a$ accounts for the others. Similarly, repeating $M_b M_d$ and $M_d M_b$ translates square 1 horizontally.
    b) $(M_b M_d)^2 (M_c M_a)^2 (1) = 73$ (8 reflections)
    c) Apply to square 213 the inverses of the reflections that mapped 1 upon 213 (that is, the same reflections in the reverse order), and then follow with the reflections mapping 1 upon 475.

13. a) 10; 6; 2; 9; 7; 1
    c) $M_c M_a M_c = M_a M_c M_a$ is the reflection in a line parallel to b and concurrent with a and c. $M_a M_b M_a = M_b M_a M_b$ is the other reflection mentioned.
    d) M_c and $M_b M_c M_b$ are the reflections in two of the sides of triangle 5. M_b,

$M_c M_a M_c$, and $M_b M_c M_b$ are the reflections in the sides of triangle 6. The reflection in the third side of triangle 5 can be obtained from these last three reflections just as reflections in the sides of triangle 10 were obtained from M_a, M_b, and M_c. The reflection in this third side of triangle 5 is

$$(M_b M_c M_b)(M_b M_a M_b)(M_b M_c M_b) = M_b M_c M_a M_c M_b$$

e) Note that $M_a M_b M_a = M_b M_a M_b$, and compare the text formula with the one given in the answer for (d). We have seen that, given the reflections in the sides of *any triangle*, we can compute the reflections in all sides of any *adjacent* triangle. And we can connect triangle 1 with any triangle in the plane through a sequence of triangles, each adjacent to its predecessor.

Exercises 12–7. None

Exercises 12–8

1. Reflecting the plane in the bisector of $\angle 1$ shows that $\angle 3 \cong \angle 4$. Under reflection in the bisector of $\angle 1$ the two "parts" of $\angle 2$ are shown to be congruent, so $\angle 2$ is also bisected.

2. Under H_m the transversal is mapped upon itself, and l is mapped upon a line parallel to l and containing a point of m. Hence $l = m$, for through each point of m the line m is the only parallel to l.

3. Under the reflection in $\overleftrightarrow{AM}$, line $\overleftrightarrow{AM}$ is mapped upon itself, and the sides of $\angle A$ are interchanged. Then B is mapped upon C and C upon B, for $\overline{AB} \cong \overline{AC}$. Hence $\angle AMB$ is mapped upon $\angle AMC$; each is a right angle, and $\overline{MC} \cong \overline{MB}$.

7. $T(B)$ lies on line $\overleftrightarrow{BC}$. But $T(B)$ must be on line $\overleftrightarrow{DC}$ since $T(A) = D$ and $T(B)$ is on the line through D parallel to $\overleftrightarrow{AB}$, namely line $\overleftrightarrow{DC}$. Hence vectors $\mathbf{v}$ and $\mathbf{v}'$ are equal.

14. a) If P is a point of the circle, each rotation about O maps P upon a point P' so that $\overline{OP} \cong \overline{OP'}$. Hence P' is on the circle.
    b) If $O \in l$ and P is any point of the circle, then $M_l(O) = O$ and $M_l(P) = P'$, $\overline{OP} \cong \overline{OP'}$ and P' is on the circle.

16. Construct B' so P bisects BB'. Draw parallels to the sides of $\angle B$ through B'. These lines intersect $\angle B$ in the points X and Y.

18. If $\overleftrightarrow{AB}$ or $\overleftrightarrow{A'B}$ is parallel to l, a special construction is required. If not, let $\overleftrightarrow{AB}$ intersect l at P, and $\overleftrightarrow{A'B}$ intersect l at P'. Now, lines $\overleftrightarrow{AP'}$ and $\overleftrightarrow{A'P}$ intersect at B'. For the case when one of $\overleftrightarrow{AB}$, $\overleftrightarrow{A'B}$ is parallel to l, the construction depends upon concepts of projective geometry and is a bit more complicated.

Chapter 13

Exercises 13–1

1. -3

2. a) b b) a c) c

3. $x < y < z$ or $z < y < x$

4. a) 6   b) 14   c) 14   d) 6

6. a) 7   b) -3 c) 3   d) -7

8. a) 9, 12   b) $-1, -8$ c) 1, 8   d) $-9, -12$

10. a) 17   b) 2   c) -13

11. c) $XY = y - x$, $YX = x - y$

Exercises 13–2

1. a) $-\frac{3}{4}$ b) $\frac{4}{3}$ c) $-\frac{6}{8}$ d) $\frac{3}{2}$ e) 0
    f) Vertical line, no slope

3. a) 5   b) $\sqrt{137}$ c) 3   d) 10

6. a) (7, 14)   b) (4, 5)   c) (1, -4)

10. a) Not collinear   b) Collinear   c) Not collinear

14. $x = 2$ 15. (0, -2)

Exercises 13–3

2. The graphs are straight lines

4. Slopes are (a) 2; (b) 3; (c) -2; (d) -3

5. a) Vertical line through $(-3, 0)$
    b) x-axis   c) y-axis

6. Slope is m; the line crosses the y-axis at $(0, d)$; $y = -(a/b)x + (-c/b)$; slope $-a/b$, crosses y-axis at $(0, -c/b)$

7. a) horizontal line   b) vertical line
    c) No points satisfy the equation
    d) Every point in the plane satisfies the equation.

8. Not both a and b equal zero.

9. If $2a - 3b + 5 = 0$, then

    $$2(a + 3) - 3(b + 2) + 5 = (2a - 3b + 5) + 6 - 6 = 0,$$

    and $(a + 3, b + 2)$ is on the line since its coordinates satisfy the equation.

Exercises 13–4

3. $5x + 3y - 2 = 0$

4. $2x - 5y + 7 = 0$

5. a) $4x - 5y + 17 = 0$ b) $4x - 5y + 20 = 0$
   c) $y = 3$ d) $x = 3$

6. $3x - 4y - 1 = 0$

7. $5x + 2y + 4 = 0$

8. a) $(-\frac{9}{2}, \frac{16}{3})$ b) $(2, -1)$
   c) $(7, 19)$ d) $(-1, 5)$

10. Lines are the same.

11. Distance is 5.

12. a) $\frac{24}{5}$ b) 2

13. Altitude $25/\sqrt{73}$; length of side $\sqrt{73}$; area $\frac{25}{2}$.

14. a) Above   b) Below

15. a) Below   b) Above

Exercises 13–5

1. $M_a(x) = T_a M_0 T_{-a}(x) = T_a M_0(x - a) = T_a(a - x) = 2a - x$

2. b) $M_a M_b(x) = M_a(-x + 2b) = -(-x + 2b) + 2a = x + 2(a - b)$
   c) $M_b M_a(x) = M_b(-x + 2a) = -(-x + 2a) + 2b = x + 2(b - a)$
   d) $T_{2b} M_a(x) = T_{2b}(-x + 2a) = -x + 2(a + b) = M_{a+b}(x)$

4. a) $x' = x + 7$ b) $x' = x + 1$ c) $x' = c - a$
      $y' = y + 3$ $y' = y + 2$ $y' = d - b$

5. a) At right angles; same distance
   b), (c) as in (a)
   d) Same distance, opposite directions
   e) As in (a)   f) As in (d).

7. a) $m(\overline{PQ}) = \sqrt{(x_1 - x_2)^2 + (y_1 - y_2)^2}$
       $m(\overline{P'Q'}) = \sqrt{(x_1 + a - x_2 - x_2 - a)^2 + (y_1 + b - y_2 - b)^2} = m(PQ)$

8. If a translation maps all the points of one figure upon all the points of a second, then the figures are congruent.

9. $T_{(3, -2)}$: $(4, 5) \rightarrow (7, 3)$, etc.

10. Yes. $T_{(-1, 3)}$ maps one upon the other.

12. No. It might be necessary to rotate one to make it coincide with the other.

14. If there is a translation $T_{(a,b)}$ which maps one vertex upon a second and also maps a third upon the fourth, the quadrilateral is a parallelogram.

15. a) Yes. Try $T_{(3,5)}$.
    b) Yes. Try $T_{(6,1)}$.
    c) No

16. $T_{(1,3)}$: $(2, 1) \to (3, 4)$. $T_{(1,3)}$: $(8, 5) \to (9, 8)$
   $T_{(-1,-3)}$: $(3, 4) \to (2, 1)$. $T_{(-1,-3)}$: $(8, 5) \to (7, 2)$
   $T_{(-6,-4)}$: $(8, 5) \to (2, 1)$. $T_{(-6,-4)}$: $(3, 4) \to (-3, 0)$
   There are three points: $(9, 8)$, $(7, 2)$, $(-3, 0)$.

Exercises 13–6
 1. a) $(2, 4)$ b) $(-1, 5)$
 2. a) $(2, 0) \to (2, -4)$, $(-1, -1) \to (-1, -3)$
    b) $(2, 0) \to (-2, 0)$, $(-1, -1) \to (1, -1)$
    c) $(2, 0) \to (2, 0)$, $(-1, -1) \to (-1, 1)$
    d) $(2, 0) \to (-8, 0)$, $(-1, -1) \to (-5, -1)$
    e) $(2, 0) \to (0, 2)$, $(-1, -1) \to (-1, -1)$
    f) $(2, 0) \to (0, -2)$, $(-1, -1) \to (1, 1)$
    g) $(2, 0) \to (-\frac{4}{13}, -\frac{20}{13})$, $(-1, -1) \to (\frac{23}{13}, \frac{11}{13})$
 3. a) A line parallel to the axis
    b) Itself
    c) The other line on P making a $30°$ angle with the axis
    d) The axis
 4. a) $(x, y) \to (2a - x, y)$ b) $(x, y) \to (-y, -x)$
    c) $(x, y) \to (x, -y)$ d) $(x, y) \to (-x, y)$

Exercises 13–7

2. $\begin{cases} x_1' = -x_1 + 2a \\ y_1' = -y_1 + 2b \end{cases}$ $\begin{cases} x_2' = -x_2 + 2a \\ y_2' = -y_2 + 2b \end{cases}$

   $(x_1' - x_2')^2 + (y_1' - y_2')^2 = (x_2 - x_1)^2 + (y_2 - y_1)^2$

3. Under reflection in $(-2, -2)$,
   $(-7, 2) \to (3, -6)$; $(-4, 1) \to (0, -5)$; $(1, 6) \to (-5, -10)$

6. $(x, y) \xrightarrow{\ M_{x=y}\ } (y, x) \xrightarrow{\ M_{y=0}\ } (y, -x) = R_{-90°}(x, y)$, etc.

7. A geometric solution is pictured:

$\begin{cases} x' = 8 + (5 - y) = 13 - y \\ y' = 5 + (x - 8) = x - 3 \end{cases}$

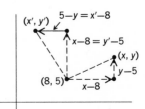

8. First reflect in $x - y + 1 = 0$ and then in $y = 3$.

$$(x, y) \xrightarrow{M_{x-y+1}} (y - 1, x + 1) \xrightarrow{M_{y=3}} (7 - y, x + 1)$$

9. a) The final image is P itself.

c) $(x, y) \to (-x + 2a_1, -y + 2b_1) \to (x - 2a_1 + 2a_2, y - 2b_1 + 2b_2)$
$\to (-x + 2a_1 - 2a_2 + 2a_3, -y + 2b_1 - 2b_2 + 2b_3)$
$\to (x + 2a_2 - 2a_3, y + 2b_2 - 2b_3)$ etc.

10. $\begin{pmatrix} 0 & 1 \\ -1 & 0 \end{pmatrix}$ is the 90° counterclockwise rotation.

11. $\begin{pmatrix} 1 & 0 \\ 0 & -1 \end{pmatrix}$ and $\begin{pmatrix} -1 & 0 \\ 0 & 1 \end{pmatrix}$ are reflections in x-axis and y-axis, respectively.

12. This is the reflection in $y = -x$.

13. $A^2 = \begin{pmatrix} -1 & 0 \\ 0 & -1 \end{pmatrix}$; $A^3 = \begin{pmatrix} 0 & 1 \\ -1 & 0 \end{pmatrix}$ is the 270° rotation.

14. $(0, 0) \to (0, 0)$; $(1, 2) \to (1, 2)$; $(3, 6) \to (3, 6)$; $(3, -4 \to (-5, 0)$;
$(4, 3) \to (0, 5)$; $(-5, 0) \to (3, -4)$; $(0, 5) \to (4, 3)$.